Handbook of Electroplating

Handbook of Electroplating

Edited by **Nathan Paul**

CLANRYE INTERNATIONAL

New Jersey

Published by Clanrye International,
55 Van Reypen Street,
Jersey City, NJ 07306, USA
www.clanryeinternational.com

Handbook of Electroplating
Edited by Nathan Paul

International Standard Book Number: 978-1-63240-269-1 (Hardback)

Printed in the United States of America.

Contents

Preface

The process of depositing a thin coating of one metal on top of a different metal is known as electroplating. This book focuses on new applications of electroplating, keeping in mind the environmental aspect and experimental design. This book provides a comprehensive overview of the new functions of electroplating. It also presents a resolution for the environmental problems caused by the electroplating process and discusses an experimental design for the optimization of electro deposition procedures. The book includes contributions by prominent experts from academics as well as electroplating industries.

All of the data presented henceforth, was collaborated in the wake of recent advancements in the field. The aim of this book is to present the diversified developments from across the globe in a comprehensible manner. The opinions expressed in each chapter belong solely to the contributing authors. Their interpretations of the topics are the integral part of this book, which I have carefully compiled for a better understanding of the readers.

At the end, I would like to thank all those who dedicated their time and efforts for the successful completion of this book. I also wish to convey my gratitude towards my friends and family who supported me at every step.

Editor

Part 1

Application of the Electroplating

Electrochemical Properties of Carbon-Supported Metal Nanoparticles Prepared by Electroplating Methods

Misoon Oh and Seok Kim
Pusan National University,
South Korea

1. Introduction

1.1 Pt and Pt-Ru catalysts for direct methanol fuel cells

There is a great demand for more potent, lightweight, efficient and reliable power-sources for a variety of transportation, portable electronics and other applications (Lamm et al., 2003). Batteries and fuel cells are alternative energy devices but batteries are only viewed as a short/mid-term option (Kuk & Wieckowski, 2005). Since transportation represents a significant portion of world energy consumption and contributes considerably to atmospheric pollution, the development of an appropriate **fuel cell** system is an important issue from both economical and environmental points of view (Arico et al., 2001). **Direct methanol fuel cells (DMFCs)** are an attractive portable power source owing to their high energy density, easy fuel handling, and a low operating temperature (Arico at al., 2000; Chen & Tang, 2002; Ren et al., 2000; Witham at al., 2003). However, DMFCs entail some serious technical obstacles. One is the relatively slow kinetics of the methanol oxidation reaction at an anode (Lima at al., 2001). Methanol oxidation reaction involves the transfer of six electrons to the electrode for complete oxidation to carbon dioxide. From a general point of view, almost all electro-oxidation reactions involving low molecular weight organic molecules, such as CO, CH_3OH, C_2H_5OH, HCOOH, HCHO, require the presence of **a Pt-based catalyst** (Arico at al., 2001). Platinum (Pt) has a high activity for methanol oxidation (Katsuaki at al., 1988, 1990; Watanabe at al., 1989). **Pt** is involved in two key steps occurring during the methanol oxidation route. One is the dehydrogenation step and the second is the chemisorption of CO (Arico at al., 2001). Pt electrocatalyst will be poisoned by intermediates of methanol oxidation, such as CO. To solve this problem, Pt was alloyed with other transition metals. Since the mid 1970s, to promote methanol electro-oxidation by Pt, the catalyst surface has been modified by the addition of a second metal to Pt (Gotz & Wedt, 1998; Hamnett at al., 1988; Mukerjee at al., 1999). The resulting Pt–Ru binary metallic catalyst is commonly accepted as the best electrocatalyst for methanol oxidation (Chu & Jiang, 2002; Gasteiger at al., 1993; Ticianelli at al., 1989; Ueda at al., 2006). In commercializing fuel cells, one of the most critical problems is the cost of metal catalysts deposited on electrodes (Chen at al., 2005; Frelink at al., 1995; Guo at al., 2005; Xiong & Manthiram, 2005; Kim at al., 2005; Kuk & Wieckowski, 2005; Qiao at al., 2005). Generally, platinum (Pt) or platinum alloy-based nanoclusters, which are impregnated on carbon supports, are the best electrocatalysts for anodic and cathodic reaction of direct methanol fuel cells (DMFCs). These catalyst materials are very expensive, and thus there is a need to minimize the catalyst loading, without sacrificing electro-catalytic activity. One way to maximize catalyst utilization is to enhance the external Pt specific surface area per unit mass of Pt. The most efficient way to achieve this goal is to reduce the size of the Pt clusters. It is reported that the particle size and distribution of Pt-based catalysts are key factors that determine their electrochemical activity and cell performance for DMFCs (Kim at al., 2006; Kim & Park, 2006, 2007; Liang at al., 2003). The activity depends also on the morphological structure of the metal–carbon composite electrode (i.e. carbon surface area, carbon aggregation, metal dispersion, and metal–carbon interaction, etc.).

1.2 Supporting materials for metal catalysts

Supporting materials are also important factor to control the size of the metal catalysts particles and their dispersion. The ideal support material should have the following characteristics: provide a high electrical conductivity, have adequate water-handling capability at the cathode, and also show good corrosion resistance under oxidizing conditions. Different substrates for catalyst particles have been tried, with the aim of improving the efficiency of methanol electro-oxidation. Generally, the electrocatalysts are supported on high-surface-area **carbon blacks** with a high mesoporous distribution and graphite character, and XC-72 carbon black is the most widely used carbon support because of its good compromise between electronic conductivity and the BET surface area. However, this conventional carbon supported catalyst offers only a low rate of methanol oxidation, partly due to low utilization of Pt, owing in turn to a low available electrochemical specific surface area for the deposition of Pt particles (Shi, 1996). Moreover, Vulcan XC-72 carbon particles generally contain sulfur groups, causing some side reaction and possibly resulting in Pt particle aggregation (Roy at al., 1996). Accordingly, considerable effort has been devoted to the development of new carbon support materials to improve both the oxidation rate and the electrode stability in methanol oxidation. Carbon supports include graphite nanofibers (GNFs), carbon nanotubes (CNTs), carbon nanocoils (CNCs), and ordered microporous carbons (OMCs) (Joo at al., 2001; Li at al., 2003; Rajesh at al., 2000; Steigerwalt at al., 2001). Novel modified carbon species have been studied as the support materials for catalyst deposition in order to obtain an available surface area, better catalyst dispersion, and the resulting high electroactivity.

Conducting polymers can be also used as suitable host matrices for dispersing metallic particles. Conducting polymer/metal-nanoparticles composites permit a facile flow of electronic charges through the polymer matrix during the electrochemical process. Additionally, electrical conducting polymers provide a low ohmic drop of electron transfer between the metal catalyst and the substrates. Also, metallic particles can be dispersed into the matrix of these polymers. By combining conducting polymers and metal particles, it is expected that novel electrodes with higher specific surface areas and enhanced electrocatalytic activities could be prepared. Recently, Y. E. Sung et al. proposed that Pt–Ru nanoparticles/electrical conducting polymer nanocomposites are effective as anode catalysts (J. H. Choi at al., 2003). At that time, the electrical conductivity of polymers was relatively low compared with amorphous carbon, and synthetic methods were limited to electrochemical polymerization from nonaqueous systems. In previous studies (Kost at al.,

1988; Lai at al., 1999), metal particles could be homogeneously dispersed on polyaniline film by constant potential electro-plating techniques. However, in our study, step-potential electroplating methods were utilized to obtain a smaller size and a better electrochemical activity.

1.3 Electrodepositon

Pulse electrodeposition has many advantages in terms of controlled particle size, stronger adhesion, uniform electrodeposition, selectivity of hydrogen, reduction of internal stress, etc. Pulse electrodeposition has three independent variables, namely, t_{on} (on-time), t_{off} (off-time) and i_p (peak current density). The properties of metal deposits can be influenced by both the on-time, during which formation of nuclei and growth of existing crystals occur, and the off-time, during which desorption of deposited ions takes place.

Electrochemical deposition occurs at cathode electrode and the deposited metal ions are exhausted along the electrodeposition. Eventually, the ion concentration at the cathode surface becomes zero. The current density at which dendrite crystals begin to form is defined as the 'limiting current density'. At this current density, the ion concentration at the deposited surface is zero. Pulse electrodeposition can raise the limiting current density because the deposited metal ions of the cathode surface can be supplied from the bulk solution during the off-time of the pulse (K. H. Choi at al., 1998).

Recently, electrochemical deposition of metal catalysts has been receiving more and more attention due to advantages such as the high purity of deposits, the simple deposition process, and the easy control of the loading mass. By applying a short-duration specific current or potential, and then repeating the process during electrodeposition, each cycle of this process can generate new metal particles (Laborde at al., 1994). Therefore, by controlling the magnitude of the current/potential and applied time, nanoparticles can be altered in size and structure. The size and dispersion of catalyst particles on the substrate determine the catalysts' performance as an electrode material.

Lee and co-workers proposed that Pt nanoparticles as catalysts for proton exchange membrane fuel cells be prepared by pulse electrodeposition (K. H. Choi at al., 1998). As they reported, a current pulse electrodeposition method resulted in better catalytic activity than did direct current (dc) electrodeposition, owing to the larger effective surface area of catalyst. In previous studies (Coutanceau at al., 2004; Kost at al., 1988; Lai at al., 1999), metal particles have successfully been homogeneously dispersed on support by constant potential electro-plating techniques.

The aim of this study was not only to overcome the problem of the complexity of catalyst preparation by using a simple electrodeposition, but also to control the size and loading level of metal particles. Furthermore, an optimal electrical signal condition, such as a certain step interval or plating time, was sought in order to enable the manufacture of a catalyst electrode offering improved catalytic activity. Besides, the effect of carbon-like species including carbon nanotubes, graphite nanofibers and conducting polymers on the electrochemical activity of metal-carbon composite electrodes will be investigated. By performing the current-voltage characterization or electrochemical methods, the possibility of fuel-cell catalyst utilization will be presented. The morphology and micro-structure of carbon-supported metal particles will be also studied and the relationship with the catalyst activity is analyzed.

2. Preparation of Pt and Pt-Ru catalysts by electrochemical deposition

2.1 Deposition of Pt on graphite nanofiber by electrochemical potential sweep method

Carbon materials used were graphite nanofibers (GNFs), which were supplied by Showa Denko Co. (Japan). These carbon fiber materials have a diameter of 100~150nm and a length of 5~50µm, resulting a large aspect ratio.

Electrochemical deposition was performed using an Autolab with a PGSTAT 30 electrochemical analysis instrument (Eco Chemie B.V.; Netherlands). The carbon materials were coated on to glassy carbon substrate with the help of 0.1% Nafion solution. A 10mM hexachloroplatinic acid (H_2PtCl_6) was dissolved in 0.5M HCl aqueous solution. A potential was swept from -700mV to -200mV (vs. Ag/AgCl) with a sweep rate of 20mV/s. A large reduction peak is shown at about –600mV (vs. Ag/AgCl). The deposited catalyst electrodes are prepared with changing the sweep times.

2.2 Deposition of Pt-Ru on carbon nanotubes by step-potential plating method

Electrodeposition of Pt-Ru nanoparticles on support was performed using an Autolab with a PGSTAT 30 electrochemical analysis instrument (Eco Chemie B.V.; Netherlands). The solid precursors, chloroplatinic acid (H_2PtCl_4, Aldrich) and ruthenium chloride ($RuCl_3$, Aldrich), were used without purification. A standard three-electrode cell was employed. Carbon nanotubes (CNTs) were selected as the support material. The CNTs, supplied from Iljin Nanotech, Korea (Multiwalled nanotubes, Purity: >99 wt.%), were used without further purification. The CNTs were synthesized by a chemical vapor deposition (CVD) process.

The CNTs mixed with 10% Nafion® perfluorosulfonated ion-exchange resin (Aldrich) solution, were dropped onto a glassy carbon electrode, selected as the working electrode. A Pt wire as the counter electrode and KCl-saturated Ag/AgCl as the reference electrode were used, respectively. Pt-Ru nanoparticles were, by the step-potential plating method, electrodeposited on the CNTs catalyst electrodes in a distilled water solution containing ruthenium chloride and chloroplatinic acid. In the deposition solution, the atomic concentration of Pt and Ru was 20 mM, respectively. A potential function generator was used to control both the step-interval time and the plating time. The potential wave form of the deposition is shown in Figure 1. The catalysts were cycled in the range of -0.3 V to -0.8 V with interval times (t_1 and t_2) of 0.03, 0.06, 0.20, and 0.50 seconds. The plating time was 20 min in all cases. However, the catalysts were prepared by changing the plating time from 6 to 12, 24, and 36 min, using constant interval times (t_1 and t_2).

2.3 Deposition of Pt-Ru on conducting polymer supports and carbon supports by step-potential plating method

2.3.1 Support materials

CBs and polyaniline were used as a support for the metal catalysts. The CBs of 24 nm average particle size, and having a DBP adsorption of 153 (cc-100g^{-1}) and a specific surface area of 112 (m^2 g^{-1}), were supplied by Korea Carbon Black Co.

Polyaniline powder was synthesized chemically by oxidative polymerization of aniline in an aqueous acidic solution (Aleshin at al., 1999; Kim & Chung, 1998; MacDiarmid & Epstein,

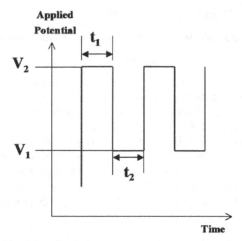

Fig. 1. Step-potential plating method showing parameters of applied potential and interval time.

1989). The aniline (8.4g, 0.09mol; Aldrich), distilled three times before use, was dissolved in 300 mL of aqueous solution containing 0.09 mol dodecylbenzene sulfonic acid (DBSA) below 5°C, and an aqueous solution (100 mL) of 0.06 mol ammonium peroxydisulfate, $(NH_4)_2S_2O_8$, was added with vigorous stirring, over a period of 30 min. The mixture was stirred continuously for 24 h. The precipitate was collected after pouring methanol and by filtration, and then washed with water and methanol three times. The resulting powder was dried under a dynamic vacuum at 40°C for 2 days. The DBSA-doped polyaniline was confirmed by measuring the electrical conductivity (~80 S/cm). The conductivity of CBs was 0.7 S/cm. PANI and CBs pellets, fabricated by compressing PANI and CBs powder under the pressure of 5.0×10^7 kg/m^2 at room temperature, were measured for conductivity using the four-probe method.

2.3.2 Electrodeposition of Pt-Ru nanoparticles on support

Electrodeposition of Pt-Ru nanoparticles on support was investigated using an Autolab with a PGSTAT 30 electrochemical analysis instrument (Eco Chemie B.V.; Netherlands). The solid precursors chloroplatinic acid (H_2PtCl_4, Aldrich) and ruthenium chloride (RuCl$_3$, Aldrich) were used without purification. A standard three-electrode cell was employed. The support materials mixed with 10% Nafion Perfluorosulfonated ion-exchange resin (Aldrich) solution was dropped onto the glassy carbon electrode as a working electrode. A Pt wire as the counter electrode and KCl-saturated Ag/AgCl as the reference electrode were used, respectively. Pt-Ru nanoparticles were, by step potential plating method, electrodeposited on the CB and polyaniline catalyst electrodes from distilled water solution containing ruthenium chloride and chloroplatinic acid. In the deposition solution, the concentration of Pt and Ru was 20 mM. A potential function generator was used to control both the step potential value and the interval time. The potential wave form of the deposition is shown in Figure 1. The catalysts were cycled in the range of -0.3 V to -0.8 V (V_1 and V_2) with an interval time (t_1 and t_2) of 0.1 seconds. The electrocatalysts were prepared by changing the plating time.

3. Characterization of Pt and Pt-Ru catalysts

3.1 Deposition of Pt on graphite nanofiber by electrochemical potential sweep method

3.1.1 Morphologies and structural properties

After Pt incorporation into carbon materials, the average crystalline sizes of Pt nanoparticles were analyzed by XRD measurements. Figure 2 shows the powder X-ray diffraction patterns of Pt catalysts deposited on GNFs as a function of sweep times. With an increase of sweep times, a sharp peak of Pt(111) at $2\theta = 39°$ and a small peak at $2\theta = 46°$ and $2\theta = 67°$ increase gradually. This gradual changes of the peak intensity can be clearly explained by the fact that and Pt was deposited successively as new particles on GNFs surface and the loading level increase with the increase of sweep times. On the other hand, it can be shown that GNFs in this study have a crystalline graphitic structure due to the strong peak at $2\theta = 26°$ and $2\theta = 54°$. All samples show the typical Pt crystalline peaks of Pt(111), Pt(200) and Pt(220).

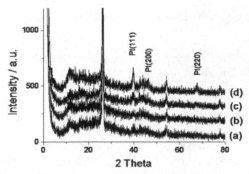

Fig. 2. Powder X-ray diffraction patterns of Pt/GNFs prepared by potential sweep method as a function of sweep times of (a) 6, (b) 12, (c) 18, and (d) 24 times.

The average size of Pt nanoparticles was calculated by using a Scherrer equation and was demonstrated in Table 1. The average size of Pt increased from 2.13nm to 8.94nm gradually by changing the sweep times of electrochemical deposition. It is interesting to note that a sharpness of peak is enhanced with an increase of sweep times.

Sample	Average size (nm)	Loading level (%)
Pt6	2.13	4
Pt12	2.90	7
Pt18	3.73	9
Pt24	8.94	11

Table 1. Average particle sizes and loading content of carbon-supported platinum catalysts as a function of sweep times.

The larger the sharpness of peaks is, the larger the average crystalline size of Pt particles is. The increase of Pt average size could be related to an aggregation of Pt nanoparticle by increasing the sweep times. It is expected that the deposited Pt particles are more aggregated and become larger. From this result, it is concluded that the particle size can be changed by controlling the condition of electrochemical deposition methods.

3.1.2 Electrochemical properties

Platinum content can be calculated by ICP-AES method. The result is summarized in Table 1. In the viewpoint of Pt content, Pt24 shows the highest value of 11%, which means the loading level increased gradually with the sweep times. Figure 3 shows the electroactivity of Pt catalyst supported on the carbon supports by linear sweep voltammograms. Peak potentials and current density were shown in Table 2. Anodic peaks for a methanol oxidation were shown at 850 ~ 900 mV for each sample. The current density of anodic peaks increased from 0.08 mA cm-2 to 0.17 mA cm-2 by increasing sweep times from 6 to 18. However, a further increase of sweep times over 18 has brought a decrease of current density. The specific current densities for each sample were also shown in Table 2. The mass activity showed the maximum value of 204 mA/mg at 18 sweep times. This value is a similar level of value (~200 mA/mg) with the previous reports.

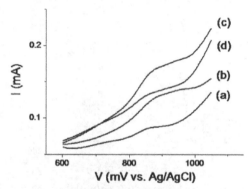

Fig. 3. I-V characteristic curves of Pt catalysts prepared by potential sweep method as a function of sweep times of (a) 6, (b) 12, (c) 18, and (d) 24 times (in 0.5M H_2SO_4 + 0.5M CH_3OH).

Sample	Peak Potential (mV)	Peak Current (μA)	Specific Activity (mA/mg)
Pt6	856	85	145
Pt12	870	125	186
Pt18	872	166	204
Pt24	860	131	120

Table 2. Peak parameters and specific activities for CV results of Pt/GNFs prepared by potential sweep method as a function of sweep times.

This means that the electrocatalytic activity has been decayed when the sweep times is over 18. Consequently, the electrocatalytic activity increased with sweep times upto 18 due to an increase of loading level of Pt. However, the electroactivity are decayed at excessive sweep times due to a large size of Pt nanocluster due to the aggregation.

Figure 4 shows impedance plots of catalysts in 0.5M H_2SO_4 + 0.5M CH_3OH. These plots were obtained by changing the frequency from 1 MHz to 0.1 Hz. The plots show an imperfect semi-circle part and a linear line part. The bulk resistance could be obtained by considering the real part of impedance at x-axis intercept. The bulk resistances were described in Figure 5. With an increase in sweep times from 6 to 18, the resistance decreased from 68 to 4 Ohm cm^2. With an increase in sweep times from 18 to 24, the resistance increased from 4 to 9 Ohm cm^2. The resistance change with sweep times showed a similar trend with the above electroactivity results. Considering together these results, the resistance change could be one of the origins for the electroactivity changes. It is thought that the resistance could be largely decreased by the improved electrical conduction between Pt particles and GNFs supports due to the enhanced loading levels.

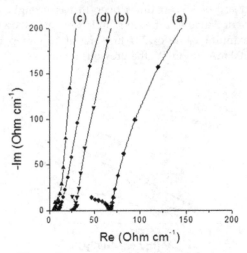

Fig. 4. Impedance plots of Pt catalysts prepared by potential sweep method as a function of sweep times of (a) 6, (b) 12, (c) 18, and (d) 24 times.

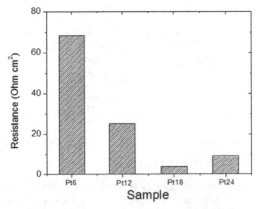

Fig. 5. Bulk resistance of Pt catalysts prepared by potential sweep method as a function of sweep times.

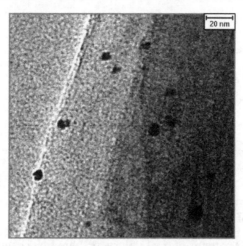

Fig. 6. TEM image of the Pt catalysts prepared by potential sweep method at 18 sweep times.

Figure 6 shows the TEM image of Pt nanoparticles deposited on GNFs supports prepared by electrochemical deposition. It has been clearly shown that 3-7nm nano-sized nanoparticles are successfully deposited and well-dispersed on GNFs surfaces. The appearance and population of particle are rather scarce due to the low loading level of 9 wt.%.

3.2 Pt-Ru on carbon nanotubes by step potential plating method

3.2.1 Size and loading level of catalysts

The crystalline structures of the Pt-Ru/CNTs catalysts were investigated by X-ray diffraction (XRD). Figure 7 shows the XRD patterns of the catalysts prepared by changing the step interval of the step-potential plating method. The peaks at $2\theta = 40°$, $47°$, $68°$, and $82°$ were associated with the (111), (200), (220), and (311) types of Pt, respectively. All of the catalysts demonstrated diffraction patterns similar to those of the Pt. In the case of the 0.03 sec interval, the catalyst showed a distinct characteristic peak and the strongest intensity. By

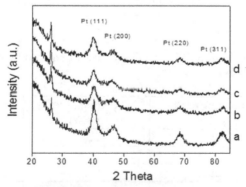

Fig. 7. X-ray diffraction patterns of PtRu/CNTs catalysts prepared by changing step interval from (a) 0.03 to (b) 0.06, (c) 0.20, and (d) 0.50 sec.

increasing the interval from 0.03 sec to 0.20 sec, the intensities of the four peaks were gradually changed. The loading contents of the catalysts were obtained separately by using ICP-AES methods, and are given in Table 3. The loading content of Pt degraded from 12.9% to 9.2%, and that of Ru changed from 7.8% to 4.7%. The weight and atomic ratio of PtRu were also described in Table 3.

Step interval (sec)	Loading of Pt (wt.%)	Loading of Ru (wt.%)	Weight ratio	Atomic ratio
0.03	12.9	7.8	1:0.61	1:1.17
0.06	11.5	6.4	1:0.56	1:1.07
0.2	10.4	5.5	1:0.53	1:1.02
0.5	9.2	4.7	1:0.51	1:0.99

Table 3. Loading level of PtRu/CNTs catalysts measured by ICP-AES method

Figure 8 shows the XRD patterns of the catalysts prepared by changing the plating time of the step-potential plating method. In the case of the 6 min plating time, the intensity of the characteristic peaks of Pt was rather small. After 24 min plating, the catalysts showed definite and enhanced peak intensities for the four kinds of peaks. The precise loading contents and PtRu ratio using the ICP-AES methods are given in Table 4. The loading content of Pt was upgraded from 4.7% to 13.1%, and that of Ru was changed from 2.4% to 8.2%.

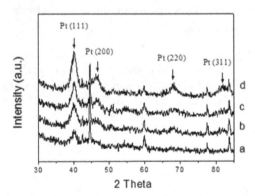

Fig. 8. X-ray diffraction patterns of PtRu/CNTs catalysts prepared by changing plating time from (a) 6 to (b) 12, (c) 24, and (d) 36 min.

Plating time (min)	Loading of Pt (wt.%)	Loading of Ru (wt.%)	Weight ratio	Atomic Ratio
6	4.7	2.4	1:0.51	1:0.98
12	8.2	4.5	1:0.55	1:1.06
24	10.4	6.1	1:0.59	1:1.13
36	13.1	8.2	1:0.63	1:1.21

Table 4. Loading level of PtRu/CNTs catalysts measured by ICP-AES method.

The average sizes of the CNTs-supported catalysts as a function of step interval and plating time are shown in Table 5 and Table 6, respectively. The average crystalline sizes were obtained by XRD measurements. The mean sizes of the particles were determined from the X-ray diffractograms, using the Scherrer equation (1) (Kinoshita, 1988)

$$L = \frac{0.9\lambda}{B_{2\theta}\cos\theta_{max}} \tag{1}$$

where λ is the X-ray wavelength (1.54056 Å for the CuKα radiation), $B_{2\theta}$ is the width of the diffraction peak at half-height, and θ_{max} is the angle at the peak maximum position.

Step interval (sec)	Crystalline size from XRD (nm)	Particle size from TEM (nm)
0.03	6.2 ± 0.2	6.4 ± 0.3
0.06	4.6 ± 0.1	4.8 ± 0.3
0.2	6.4 ± 0.2	6.8 ± 0.4
0.5	8.2 ± 0.3	9.1 ± 0.5

Table 5. Mean particle size of PtRu/CNTs catalysts obtained by XRD and TEM methods.

Plating time (min)	Crystalline size from XRD (nm)	Particle size from TEM (nm)
6	7.4 ± 0.2	7.9 ± 0.5
12	6.3 ± 0.2	6.7 ± 0.4
24	4.1 ± 0.1	4.3 ± 0.2
36	11.2 ± 0.4	11.5 ± 0.5

Table 6. Mean particle size of PtRu/CNTs catalysts obtained by XRD and TEM methods.

Particle sizes were also listed by TEM methods. In the case of the changed step intervals, the smallest particle size was obtained with the 0.06 sec interval. By increasing the interval time to 0.50 sec, the particle size was gradually increased, reflecting the growth of the deposited particles. It was concluded that the 0.06 sec interval is the best condition for obtaining the smallest particle size. Therefore, the 0.06 sec interval was used to study the effect of changed plating time on the catalytic activity.

The average catalyst nanoparticle size was 7.9 nm at 6 min plating time. It was considered that particle nucleation is not as efficient as particle growth at the initial stage of electrodeposition. After 24 min plating, the particle nucleation was efficient enough to produce a new generation of small particles, resulting in the decrease of the average particle size. Although more precise nucleation and growth mechanisms are necessary for Pt-Ru nanoparticles, it was found that the smaller particle size could be obtained after an initial activation state. The smallest nanoparticles of 4.3 nm were obtained by electrodeposition after 24 min plating time. It was concluded that this is the optimal plating time for obtaining the smallest particle catalysts.

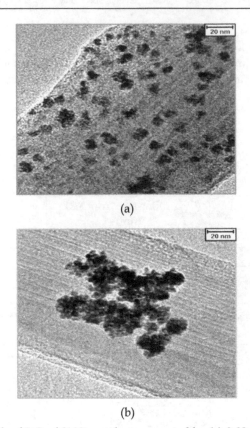

(a)

(b)

Fig. 9. TEM micrograph of PtRu/CNTs catalysts prepared by (a) 0.03 sec and (b) 0.5 sec step intervals.

The particle sizes and morphologies of the CNTs-supported Pt-Ru catalysts were investigated by TEM. Figure 9 shows TEM images of nanoparticle catalysts that were prepared on CNTs by electrodeposition at 0.06 sec and 0.50 sec interval times, which showed the smallest and the largest particle size, respectively. In the case of (a) the 0.06 sec interval, the image shows well-dispersed 3-6 nm nanoparticles on the surface of the CNTs. In contrast to this, in the case of (b) the 0.50 sec interval, the image shows rather aggregated catalyst particles of 5 – 8 nm size. It could be concluded that with increased the step-interval time, particles tend to increase in size, and also, to an extent, to aggregate. The TEM images of catalysts prepared by different plating times and a constant 0.06 sec step interval, are also shown in Figure 10. In the case of (a) the 6 min plating time, the population of the deposited particles, arranged in nanoclusters with individual particle sizes ranging from 5 to 7 nm, is rather low, reflecting the low loading content at the early state of deposition. In the case of (b) the 24 min plating time, the population of the particles, ranging in size from 3 to 5 nm, was enhanced, manifesting the increased loading content. The average particle size, as can be seen, was decreased in the case of the 24 min plating time. And the average crystalline sizes calculated from the XRD peak widths were found to be fairly consistent with particle sizes from the TEM results, as shown in Table 5 and Table 6.

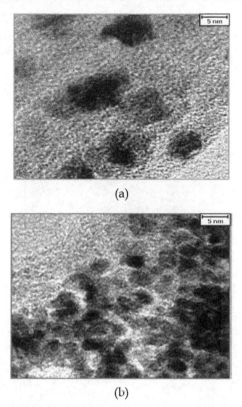

(a)

(b)

Fig. 10. TEM micrograph of PtRu/CNTs catalysts prepared with (a) 6 min and (b) 24 min plating times.

3.2.2 Electrochemical properties of catalysts

The electrochemical properties of the catalysts were investigated by linear sweep voltammetry in 1 M CH_3OH + 0.5 M H_2SO_4 aqueous solution. Figure 11 shows the current-voltage curves of the CNTs-supported catalysts, representing the electrochemical behavior of methanol oxidation. The electrochemical activities of catalysts prepared by changed step intervals were studied. In the case of the 0.03 sec interval, the catalyst curve shows a slight oxidation peak at 780 mV and an increasing current density. In the case of the 0.06 sec interval, the curve shows a slight oxidation peak at 740 mV and, again, an increasing current density. The oxidation peak was shifted to the low potential, indicating the more feasible oxidation reaction of methanol. Furthermore, the current density was higher than in the case of the 0.03 sec interval. In the cases of 0.20 sec and 0.50 sec intervals, in contrast, the oxidation peak was not clearly shown, and the current density of the catalyst was degraded. These results indicate that the electro-plating step intervals over 0.20 sec can have a negative influence on the electrochemical activity of deposited catalysts.

Figure 12 shows the current-voltage curves of the CNTs-supported catalysts prepared by changed step intervals. At the early stage of 6 and 12 min plating, the current density at the same potential was increased gradually with the plating time, meaning that the catalytic activity was increased by the enhanced loading contents. In the case of 24 min plating, the curve shows a slight oxidation peak at 730 mV, a different phenomenon from those for the other samples. Furthermore, an onset potential, which is the potential of oxidation current to begin to rise, was shifted to the lower potential of ~500 mV. The onset potentials for the other samples, were ~800 mV. It was concluded, accordingly, that the electrochemical activity was highly improved in this case of 24 min plating, which advance was related to the fact that smaller particles and higher dispersion of catalysts result in a larger available catalyst surface area and better electrocatalytic properties for methanol oxidation.

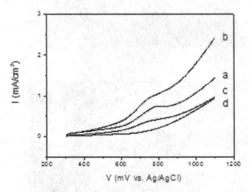

Fig. 11. Current-voltage curves of PtRu/CNTs catalysts prepared by changing step interval from (a) 0.03 to (b) 0.06, (c) 0.20, and (d) 0.50 sec.

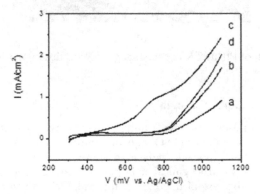

Fig. 12. Current-voltage curves of PtRu/CNTs catalysts prepared by changing plating time from (a) 6 to (b) 12, (c) 24, and (d) 36 min.

The electrochemical activity increased with increasing plating time, reaching the maximum at 24 min, and then decreased at 36 min. However, the catalysts showed an increased Pt content, in proportion to the plating time, to 36 min. That is, the Pt content was the highest when the plating time was 36 min, even though the optimal plating time, as we saw, was 24min. This puzzling result was considered to have originated in the catalyst's smallest, 4.3 nm particle size and highest specific surface area. These electroactivity changes as a function of electrical condition were inversely proportional to the size of the nanoparticle catalysts, strongly indicating that the higher electroactivity could have been enabled by the decreasing average size of nanoparticle catalysts, resulting in the increase of the efficient specific surface area for an electro-catalytic reaction.

Figure 13 shows the weight-based current densities of the prepared catalysts, according to the above conditions. At the 0.06 sec step-interval, as shown by graph (A), the specific current densities of the catalysts showed the highest value, 160 (mA/mg). At step intervals over 0.06 sec, the current density was degraded, increasingly with each interval. When the plating time was increased to 24 min, the specific current density of catalysts, as shown by graph (B), was gradually enhanced to 202 (mA/mg). However, the 36 min plating time showed a decreased value of deposited catalyst. Therefore, it could be deduced that the electroactivity of catalysts prepared by electrodeposition is strong dependent on the electrical signal condition manifest in the step interval and the plating time.

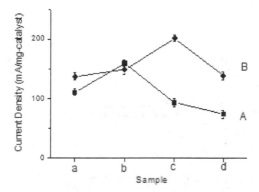

Fig. 13. Specific current density of PtRu/CNTs catalysts prepared (A) by changing step interval from (a) 0.03 to (b) 0.06, (c) 0.20, and (d) 0.50 sec, or (B) by changing plating time from (a) 6 to (b) 12, (c) 24, and (d) 36 min.

To check the specific surface area of catalysts, cyclic voltammograms (CVs) had been perfored. Figure 14 shows the CVs of the prepared catalysts in 1.0M sulphuric acid solution and the calculated electrochemical surface area are listed in Table 7. Catalyst by 0.06 sec interval time showed the highest specific surface area. In the case of changing the plating time, the specific surface area was increased upto 24 min plating time and then was decreased. Over 24 min plating, the particle was thought to be overlapped causing the decrease of the specific surface area and slight increase in particle size. This result supported that the catalytic activity was strongly dependent on the particle size and specific surface area.

3.3 Pt-Ru on conducting polymer supports and carbon supports by step potential plating method

3.3.1 Size and loading level of catalysts

The crystalline structures of the Pt-Ru/CBs and Pt-Ru/polyaniline catalysts were investigated by X-ray diffraction (XRD). Figure 15 and Figure 16 show the XRD patterns of the catalysts prepared by changing the plating time of the step-potential plating method. The peaks at $2\theta = 40°, 47°, 68°$, and $82°$ were associated with the (111), (200), (220), and (311)

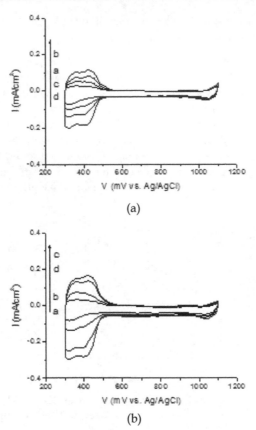

(a)

(b)

Fig. 14. Cyclic voltammograms of PtRu/CNTs catalysts prepared (A) by changing step interval from (a) 0.03 to (b) 0.06, (c) 0.20, and (d) 0.50 sec and (B) by changing plating time from (a) 6 to (b) 12, (c) 24, and (d) 36 min.

(a)	Step interval (sec)	Electrochemical surface area (cm²)	Specific surface (m²/g)
	0.3	3.21	63.2
	0.6	4.26	89.5
	0.2	2.16	51.7
	0.5	1.21	43.3
(b)	Plating time (sec)	Elctrochemical surface area (cm²)	Specific surface area (m²/g)
	6	1.97	71.8
	12	3.53	83.2
	24	6.83	98.2
	36	6.14	68.2

Table 7. Surface area of catalyst by changing (a) step interval and (b) plating time that was calculated from the H_2 absorption peak in CVs (Figure 14).

types respectively. All of the catalysts demonstrated diffraction patterns similar to those of the Pt. The characteristic peaks for Ru were not clearly shown in the XRD patterns. In the case of the 6 min plating time, the characteristic peaks of Pt were not distinct, indicating inefficient electrodeposition of the metal catalysts. For the 12 min plating time, Pt(111) appeared at ~40°. After 24 min plating, the catalysts showed definite and enhanced peak intensity for the four kinds of peaks. In the case of 36 min plating, the catalysts clearly showed the characteristic peaks. The precise loading contents of the catalysts were obtained by using ICP-AES methods, and are given in Table 8. The loading content of Pt was upgraded from 2.1% to 7.2%, and that of Ru was changed from 1.0% to 3.1%. The loading content had been increased proportionally as a function of plating time.

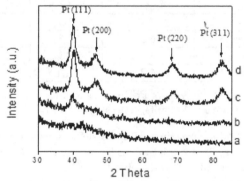

Fig. 15. X-ray diffraction patterns of Pt-Ru/CBs catalysts prepared for different plating times of (a) 6, (b) 12, (c) 24, and (d) 36 min.

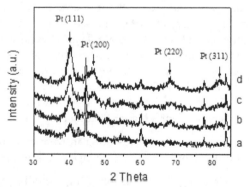

Fig. 16. X-ray diffraction patterns of Pt-Ru/polyaniline catalysts prepared for different plating times of (a) 6, (b) 12, (c) 24, and (d) 36 min.

Similar to the Figure 15 results, and as shown in Figure 16, the crystalline peaks of the polyaniline-supported catalysts were not clearly evident for the initial plating time of 6 min, except for the appearance of a small Pt (111) peak. After 12 min plating, the catalysts showed the four characteristic peaks. With increasing plating time after 24 min, the peaks became sharp and definite.

Plating time (min)	Crystalline size (nm)[a]	Particle size (nm)[b]	Pt (wt.%)[c]	Ru (wt.%)[c]	Alloyed Ru (%)
6	-	8.1	2.1	1.0	21
12	5.2± 0.3	5.4	3.9	1.2	24
24	3.4± 0.2	3.6	6.1	2.1	32
36	4.2± 0.2	4.6	7.2	3.1	26

Table 8. Mean size and loading contents of Pt-Ru/CBs catalysts.; a: measured from XRD results, b: measured from TEM results, c: measured from ICP-AES results.

Plating time (min)	Crystalline size (nm)[a]	Particle size (nm)[b]	Pt (wt.%)[c]	Ru (wt.%)[c]	Alloyed Ru (%)
6	-	7.4	2.3	1.3	20
12	4.3± 0.3	4.5	7.6	3.1	25
24	2.9± 0.2	3.1	9.3	4.6	35
36	3.9± 0.3	4.2	11.1	5.3	27

Table 9. Mean size and loading contents of Pt-Ru/polyaniline catalysts.; a: measured from XRD results, b: measured from TEM results, c: measured from ICP-AES results.

Considering particle crystalline size, the average sizes of CBs-supported catalysts showed the smallest value, 3.4 nm, at 24 min plating, as shown in Table 1. Beside, particle sizes by TEM results were shown in Table 8 and Table 9. In the case of 6 min plating time, the average size was ~8 nm. It was considered that particle nucleation was not as efficient as the particle growth at the initial stage of electrodeposition. After 24 min plating, the particle nucleation was efficient enough to produce a new generation of small particles, resulting in the decrease of the average crystalline size. Although more precise nucleation and growth mechanisms are necessary for Pt-Ru nanoparticles, it was found that the smaller crystalline size could be obtained after an initial activation state. This behavior was also observed in the case of polyaniline supports, as shown in Table 9. Regardless of the support materials, the smallest nanoparticles were obtained by electrodeposition after 24 min plating time. The loading content of Pt or Ru was 11.1% or 5.3%, respectively, after 36 min plating time. These values were slightly higher that those of the CBs supports. It was concluded that conducting polymer supports are more beneficial for a higher loading for electrodeposition of catalysts.

The peak position of PtRu catalysts had been shifted in comparison with that of pure Pt catalyst. The shift of peak position could mean the change of a lattice parameter.

We had calculated the alloying degree by the following formula (Antoline & Cardellini, 2001);

$$l_{PtRu} = 0.3916 - 0.124 x_{Ru} \qquad (2)$$

(where l_{PtRu} is the lattice parameter of PtRu catalysts, and x_{Ru} is the percent of Ru in the alloy) The lattic parameter of PtRu is smaller than that of Pt, meaning a part of Ru had entered into the crystal lattice of Pt. The alloying degree of Ru had been inserted in Table 8 and 9.

The particle sizes and morphologies of the Pt-Ru/CBs and Pt-Ru/polyaniline catalysts were investigated by TEM. Figure 17 shows a TEM image of nanoparticle catalysts that were prepared on CBs by electrodeposition with plating time. 24 min plating showed the smallest particle sizes. This shows the nanoparticles in the 2.5-5.0 nm size range. Figure 18 shows a TEM image of catalysts that were prepared on polyaniline supports with changing plating time. 24 min plating shows nanoclusters with individual particles of 2.5-4.1 nm size. The average crystalline sizes calculated from the XRD peak widths were found to be fairly consistent with those from the TEM results, as shown in Table 8 and Table 9.

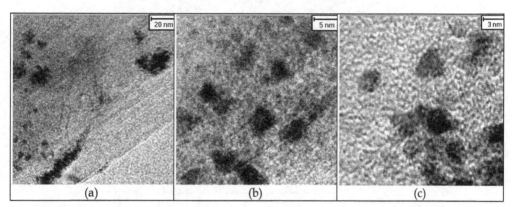

Fig. 17. TEM micrograph of Pt-Ru/CBs catalysts by (a) 6 min, (b) 12 min, and (c) 24 min plating time.

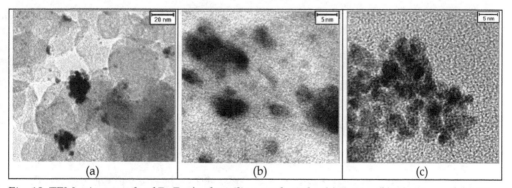

Fig. 18. TEM micrograph of Pt-Ru/polyaniline catalysts by (a) 6 min, (b) 12 min, and (c) 24 min plating time.

3.3.2 Electrochemical properties of catalysts

The electrochemical properties of the catalysts were investigated by cyclic voltammetry in 1 M CH_3OH + 0.5 M H_2SO_4 aqueous solution. Figure 19 shows the representative current-voltage curves of the CBs-supported catalysts, presenting the electrochemical behavior of methanol oxidation. Voltammetric behavior depends on the Pt content. The electrochemical activity increased with increasing plating time, reaching the maximum at 24 min, and then slightly decreased. However, the catalysts showed an increased Pt content with plating time to 36 min. The optimal plating time was 24min, although the Pt content was the highest when the plating time was 36 min. The catalyst by 24 min plating showed the highest current density for methanol oxidation, indicating the highest electroactivity by an enhanced specific surface area of a reaction site for metallic catalysts. This result was considered to have been originated from the catalyst's smaller particle size and lower aggregation.

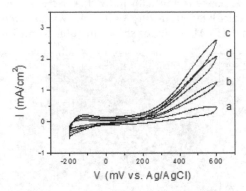

Fig. 19. Cyclic voltammograms of Pt-Ru/CBs catalysts prepared for different plating times of (a) 6, (b) 12, (c) 24, and (d) 36 min in 1 M methanol solution (scan rate: 20mV/s).

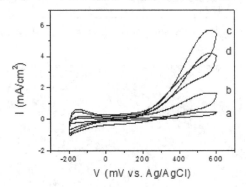

Fig. 20. Cyclic voltammograms of Pt-Ru/polyaniline catalysts prepared for different plating times of (a) 6, (b) 12, (c) 24, and (d) 36 min in 1 M methanol solution (scan rate: 20mV/s).

In the case of the polyaniline-supported catalysts, the cyclic voltammograms shown in Figure 20 also exhibit methanol oxidation peak. The catalyst deposited on PANI showed a rather definite oxidation peak at ~520 mV. Similar to the CBs supports, the catalysts showed the highest electroactivity at 24 min plating. Indeed, the catalysts by 24 min plating exhibited an average size of 3.1 nm, whereas the catalysts by 36 min plating showed an average size of 4.2 nm. Smaller particles of catalysts might result in a large available catalyst surface area and good electrocatalytic properties for methanol oxidation.

To determine the carbon and PANI support influences, individually, on the oxidation current of catalysts, supports without metal catalysts were studied in 1 M CH_3OH + 0.5 M H_2SO_4 aqueous solution, as shown in Figure 21. The carbon support did not show any electrochemical activity except for some small capacitive current. By contrast, the PANI support showed a definite cathodic/anodic peak, indicating an oxidation/reduction reaction (Aleshin at al., 1999; Kim & Chung, 1998; MacDiarmid & Epstein, 1989). However, it was concluded that polyaniline supports could not function alone as catalysts for methanol oxidation.

To check the specific surface area of catalysts, cyclic voltammograms (CVs) had been performed. Figure 22 shows the CVs of the PANI- or CBs-supported catalysts in 1.0M sulphuric acid solution. H_2 adsorption/desorption peak were observed at –200 mV for both case and the electrochemical surface area are calculated (PtRu/PANI: 4.3 cm^2, PtRu/CBs: 7.4 cm^2). By considering the different metal loading, we obtained the specific surface area by dividing the metal weight (PtRu/PANI: 62 m^2/g, PtRu/CBs: 75 m^2/g). Catalyst deposited on PANI showed the higher specific surface area than PtRu/CBs. This result supported that the effective surface area could be dependent on the particle size and aggregation degree.

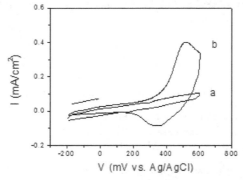

Fig. 21. Cyclic voltammograms of (a) CBs and (b) polyaniline supports in 1 M methanol solution (scan rate: 20mV/s).

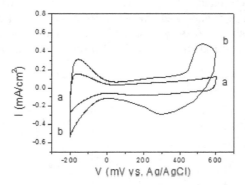

Fig. 22. Cyclic voltammograms of (a) Pt-Ru/CBs and (b) Pt-Ru/polyaniline prepared by 24 min plating in 1 M sulfuric acid solution (scan rate: 20mV/s).

Beside, chronoamperometry could be used to obtain the apparent diffusion coefficient of ions in electrochemical reactions. The current responses with time were shown in Figure 23. After the potential was raised abruptly from -0.2 to 0.6V, the current response was recorded. Using a following equation, the apparent diffusion coefficient had been calculated (Antoline & Cardellini, 2001; Bard & Faulkner, 1980).

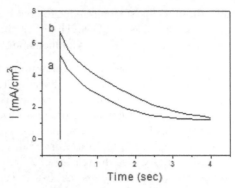

Fig. 23. Chronoamperometry results of (a) Pt-Ru/CBs and (b) Pt-Ru/polyaniline by 24 min plating in 1 M methanol solution.

$$\ln\left(\frac{i}{i_O}\right) = -\frac{\pi^2 D}{l^2} t \tag{3}$$

where i_o is the initial current, i the current, l sample thickness and t the time.

The obtained diffusion coefficients were like following; PtRu/PANI: 6.1×10^{-9} cm/s, PtRu/CBs: 4.2×10^{-9} cm/s. The former case was a higher value than the latter case. This result could be one of the reasons of the improved electroactivity.

Figure 24 shows the specific current densities of the prepared Pt-Ru/CBs and Pt-Ru/polyaniline catalysts. The current densities of the different materials-supported catalysts as a function of plating time showed similar behaviours. The PANI-supported catalysts showed enhanced electroactivity compared with the carbon-supported catalysts. In the early stage of 6 min plating, the difference of enhanced current density (41-9 = 32 (mA/mg)) was almost similar to the current density (27 (mA/mg)) of PANI itself. However, in the case of 24 min plating, the difference of increased current density (121-68 = 53 (mA/mg)) was much higher

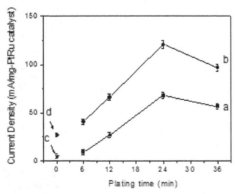

Fig. 24. Specific current density for oxidation peaks of (a) Pt-Ru/CBs and (b) Pt-Ru/polyaniline catalysts prepared for different plating times ((c) CBs and (d) polyaniline supports themselves were added for comparison) (at 600 mV).

than the current density of PANI itself. Accordingly, it was concluded that the improved electroactivity of the PANI-supported catalysts was a result not only of the combined activity of the catalysts and the PANI; one of the main sources of the improved electroactivity was the high electronic conductivity of DBSA-doped polyaniline (~80 S/cm) compared with that of the CBs (~0.7 S/cm) (Aleshin at al., 1999). Additionally, we had found that the electrochemical area of catalysts had been increased. The increased electrochemical area might be related to the small particle size and low degree of aggregation.

4. Conclusions

The electrochemical deposition and characterization of Pt and Pt-Ru nanoparticles on carbons and conducting polymer supports were investigated. Pt and Pt-Ru particles were successfully electrodeposited by electrochemical potential sweep and step-potential plating methods.

At the electrochemical potential sweep method, the average size and loading level of Pt particles increased gradually with increase of potential sweep times. The electroactivity of Pt catalyst electrode showed a highest performance at 18 sweep times due to the best particle size and loading level. It was thought that excessive sweep time brought the decay of electroactivity due to the larger particle size and degraded particle dispersion. It was concluded that the smallest particle size and the smallest bulk resistance could influence the improved activity.

The particle size and loading level of Pt-Ru was also controlled by step-potential plating method. It was found that a smaller interval time under same plating time enabled a higher loading content of electro-deposited metal particles. By contrast, Pt loading content was enhanced with the increase of plating time. With increased step interval time, particles tended to increase in size, and also, to an extent, to aggregate. These results correlated to the fact that smaller particles and higher available catalyst surface area could bring the better electrocatalytic properties for methanol oxidation. The specific surface area was confirmed by measuring the H_2 absorption peak. This result supported that the catalytic activity was strongly dependent on specific surface area.

Finally, the properties of the Pt-Ru catalysts were affected by support materials. Pt-Ru/CBs showed 68mA/mg, Pt-Ru/polyaniline showed 121mA/mg, and Pt-Ru/CNTs showed 202mA/mg for values of methanol oxidation specific current. The electroplating method and the physic-chemical feature of support materials affected the deposited particle size of Pt-Ru catalysts and tendency of aggregation.

5. Acknowledgment

This research was supported by the Converging Research Center Program through the Ministry of Education, Science and Technology (Grant No.: 2011K000643), the Ministry of Knowledge and Economy (Material source Technology Project Grant No. 10037238.2011), and Basic Science Research Program through the National Research Foundation of Korea (NRF) funded by the Ministry of Education, Science and Technology (Grant No.: 2011-0009007).

6. References

Aleshin, A.N.; Lee, K.; Lee, J.Y.; Kim, D.Y. & Kim, C.Y. Comparison of electronic transport properties of soluble polypyrrole and soluble polyaniline doped with dodecylbenzene-sulfonic acid. *Synth. Met.*, Vol.99, No.1, (January 1999), pp. 27-33.

Antoline, E. & Cardellini, F. Formation of carbon supported PtRu alloys: an XRD analysis. *J. Alloys Compd.*, Vol.315, No.1-2, (February 2001), pp. 118-122.

Arico, A.S.; Creti, P.; Modica, E.; Monforte, G.; Baglio, V. & Antonucci, V. Investigation of direct methanol fuel cells based on unsupported Pt–Ru anode catalysts with different chemical properties. *Electrochim. Acta*, Vol.45, No.25-26, (August 2000), pp. 4319-4328.

Arico, A.S.; Srinivasan, S. & Antonucci, V. DMFCs: from fundamental aspects to technology development. *Fuel Cells*, Vol.1, No.1 (September 2001), p. 133-161.

Bard, A.J. & Faulkner, L.R. Electrochemical Method, *Wiley*, New York, 1980 Chapter. 3,6, and 10.

Chen, C.Y. & Tang, P. Performance of an air-breathing direct methanol fuel cell. *J. Power Sources*, Vol.123, No.1, (September 2003), pp. 37-42.

Chen, C.Y.; Yang, P.; Lee, Y.S. & Lin, K.F. Fabrication of electrocatalyst layers for direct methanol fuel cells. *J. Power Sources*. Vol.141, No.1, (February 2005), pp. 24-29.

Choi, J.H.; Park, K.Y.; Kim, Y.M.; Lee, J.S. & Sung, Y.E. Nano-composite of PtRu alloy electrocatalyst and electronically conducting polymer for use as the anode in a direct methanol fuel cell. *Electrochim. Acta*. Vol.48, No.9, (August 2003), pp. 2781-2789.

Choi, K.H.; Kim, H.S. & Lee, T.H. Electrode fabrication for proton exchange membrane fuel cells by pulse electrodeposition. *J. Power Sources*. Vol.75m No.2, (October 1998), pp. 230-235.

Chu, D. & Jiang, R. Novel electrocatalysts for direct methanol fuel cells. *Solid State Ionics*, Vol.148, No.3-4, (June 2002), pp. 591-599.

Coutanceau, C.; Rakotondrainibe, A.; Lima, A.; Garnier, E.; Pronier, S.; Leger, J.M. & Lamy, C. Preparation of Pt-Ru bimetallic anodes by galvanostatic pulse electrodeposition: Characterization and application to the direct methanol fuel cell. *J. Appl. Electrochem.*, Vol.34, No.1, (January 2004), pp. 61-65.

Frelink,T.; Visscher, W. & Rvan, J.A. Particle size effect of carbon-supported platinum catalysts for the electrooxidation of methanol. *J. Electroanal. Chem.*, Vol. 382, No.1-2, (February 1995), pp. 65-72.

Gasteiger, H.A.; Markovic, N.; Ross, P.N. & Cairns, E.J. Methanol electrooxidation on well-characterized Pt-Ru alloys. *J. Phys. Chem.*, Vol.97, No.46, (1993), pp. 12020-12029.

Gotz, M. & Wendt, H. Binary and ternary anode catalyst formulations including the elements W, Sn and Mo for PEMFCs operated on methanol or reformate gas. *Electrochim. Acta*, Vol.43, No.24, (August 1998), pp. 3637-3644.

Guo, J.W.; Zhao, T.S.; Prabhuram, J. & Wong, C.W. Preparation and the physical/electrochemical properties of a Pt/C nanocatalyst stabilized by citric acid for polymer electrolyte fuel cells. *Electrochim. Acta*. Vol.50, No.10, (March 2005), pp. 1973-1983.

Hamnett, A.; Kenndey, B.J. & Weeks, S.A. Base metal oxides as promotors for the electrochemical oxidation of methanol. *J. Electroanal. Chem.*, Vol.240, No.1-2, (January 1988), pp. 349-353.

Joo, S.H.; Choi, S.J.; Oh, I.; Kwak, J.; Liu, Z.; Terasaki, O. & Ryoo, R. Ordered nanoporous arrays of carbon supporting high dispersions of platinum nanoparticles. *Nature*. Vol.412, No.6843, (July 2001), pp. 169-172.

Katsuaki, S.; Kohei, U.; Hideaki, K. & Yoshinobu, N. Structure of Pt microparticles dispersed electrochemically onto glassy carbon electrodes: Examination with the scanning tunneling microscope and the scanning electron microscope. *J. Electroanal. Chem.*, Vol.256, No.9, (December 1988), pp. 481-487.

Katsuaki, S.; Ryuhei, I. & Hideaki, K. Enhancement of the catalytic activity of Pt microparticles dispersed in Nafion-coated electrodes for the oxidation of methanol by RF-plasma treatment. *J. Electroanal. Chem.*, Vol.284, No.2, (May 1990), pp. 523-529.

Kim, S. & Chung, I.J. Annealing effect on the electrochemical property of polyaniline complexed with various acids. *Synth. Met.*, Vol.97, No.2, (September 1998), pp. 127-133.

Kim, S.; Cho, M.H.; Lee, J.R.; Ryu, H.J. & Park, S.J. Electrochemical Behaviors of Platinum Catalysts Deposited on the Plasma Treated Carbon Blacks Supports. Kor. Chem. Eng. Res. Vol.43, (2005), pp. 756-760.

Kim, S.; Cho, M.H.; Lee, J.R. & Park, S.J. Influence of plasma treatment of carbon blacks on electrochemical activity of Pt/carbon blacks catalysts for DMFCs. *J. Power Sources*. Vol.159, No.1, (September 2006), pp. 46-48.

Kim, S. & Park, S.J. Effects of chemical treatment of carbon supports on electrochemical behaviors for platinum catalysts of fuel cells. *J. Power Sources*. Vol.159, No.1, (September 2006), pp. 42-45.

Kim, S. & Park, S.J. Effect of acid/base treatment to carbon blacks on preparation of carbon-supported platinum nanoclusters. *Electrochim. Acta*. Vol.52, No.9, (February 2007), pp. 3013-3021.

Kinoshita, K. Carbon: Electrochemical and Physicochemical Properties, *John Wiley*, New York, 1988, p. 31.

Kost, K.M.; Bartak, D.E.; Kazee, B. & Kuwana, T. Electrodeposition of platinum microparticles into polyaniline films with electrocatalytic applications. *Anal. Chem.*, Vol.60, No.21, (1988), pp. 2379-2384.

Kuk, S.T. & Wieckowski, A. Methanol electrooxidation on platinum spontaneously deposited on unsupported and carbon-supported ruthenium nanoparticles. *J. Power Sources*, Vol.141, No.1, (February 2005), pp. 1-7.

Laborde, H.; Leger, J.M.; & Lamy, C. Electrocatalytic oxidation of methanol and C1 molecules on highly dispersed electrodes Part II: Platinum-ruthenium in polyaniline. *J. Appl. Electrochem.*, Vol.24, No.10, (October 1994), pp. 1019-1027.

Lai, E.K.W.; Beattie, P.D.; Orfino, F.P.; Simon, E. & Holdcroft, S. Electrochemical oxygen reduction at composite films of Nafion®, polyaniline and Pt. *Electrochim. Acta*. Vol.44, No.15, (1999), pp. 2559-2569.

Lamm, A.; Gasteiger, H.; Vielstich, W. (Eds.), (2003). *Handbook of Fuel Cells*, Wiley-VCH, Chester, UK.

Li, W.; Liang, C.; Qiu, J.; Zhou, W.; Qiu, J.; Zhou, Z.; Sun, G. & Xin, Q. Preparation and Characterization of Multiwalled Carbon Nanotube-Supported Platinum for Cathode Catalysts of Direct Methanol Fuel Cells. *J. Phys.Chem. B*. Vol.107, No.26, (2003), pp. 6292-6299.

Lima, A.; Cutanceau, C.; Leger, J.M. & Lamy, C. Investigation of ternary catalysts for methanol electrooxidation. *J. Appl. Electrochem.*, Vol.31, No.4, (April 2001), pp. 379-386.

MacDiarmid, A.G. & Epstein, A.J. Polyanilines: A novel class of conducting polymers. *Faraday Discuss. Chem. Soc.*, Vol.85, (1989), pp. 317-332.

Mukerjee, S.; Lee, S.J.; Ticianelli, E.A.; Mcbreen, J.; Grgur, B.N.; Markovic, N.M.; Ross, R.N.; Giallombardo, J.R. & Castro, De E.S. Investigation of enhanced CO tolerance in proton exchange membrane fuel cells by carbon supported PtMo alloy catalyst. *Electrochem. Solid-State Lett.*, Vol.2, No.1, (1999), pp. 12-15.

Qiao, H.; Kunimatsu, M. & Okada, T. Pt catalyst configuration by a new plating process for a micro tubular DMFC cathode. *J. Power Sources.* Vol.139, No.1-2, (January 2005), pp. 30-34.

Rajesh, B.; Ravindranathan, T.K.; Bonard, J.M. & Viswanathan, B. Preparation of a Pt–Ru bimetallic systemsupported on carbon nanotubes. *J. Mater. Chem.*, Vol.10, (June 2000), pp. 1757-1759.

Ren, X.; Zelenay, P.; Thomas, S.; Davey, J. & Gottesfeld, S. Recent advances in direct methanol fuel cells at Los Alamos National Laboratory. *J. Power Sources*, Vol.86, No.1-2, (March 2000), pp. 111-116.

Roy, S.C.; Christensen, P.A.; Hamnett, A.; Thomas, K.M. & Trapp, V. Direct Methanol Fuel Cell Cathodes with Sulfur and Nitrogen-Based Carbon Functionality. *J. Electrochem. Soc.*, Vol.143, No.10, (October 1996), pp. 3073-3079.

Shi, H. Activated carbons and double layer capacitance. *Electrochim. Acta.* Vol.41, No.10, (June 1996), pp. 1633-1639.

Steigerwalt, E.S.; Deluga, G.A.; Cliffel, D.E. & Lukehart, C.M. A Pt-Ru/graphitic carbon nanofiber nanocomposite exhibiting high relative performance as a direct-methanol fuel cell anode catalyst. *J. Phys. Chem. B*, Vol.105, No.34, (August 2001), pp. 8097-8101.

Ticianelli, E.; Beery, J.G.; Paffett, M.T. & Gottesfeld, S. An electrochemical, ellipsometric, and surface science investigation of the PtRu bulk alloy surface. *J. Electroanal. Chem.*, Vol.258, No.1, (January 1989), pp. 61-77.

Ueda, S.; Eguchi, M.; Uno, K.; Tsutsumi, Y. & Ogawa, N. Electrochemical characteristics of direct dimethyl ether fuel cells. *Solid State Ionics.* Vol.177, No.19-25, (October 2006), pp. 2175-2178.

Watanabe, M.; Saeguae, S. & Stonelhart, P. High platinum electrocatalyst utilizations for direct methanol oxidation. J. Electroanal.Chem., Vol.271, No.1-2, (June 1989), pp. 213-220.

Witham, C.K.; Chun, W.; Valdez, T.I. & Narayanan, S.R. Performance of direct methanol fuel cells with sputter-deposited anode catalyst layers. *Electrochem. Solid-State Lett.* Vol.3, No.11, (November 2000), pp. 497-500.

Xiong, L. & Manthiram, A. Catalytic activity of Pt–Ru alloys synthesized by a microemulsion method in direct methanol fuel cells. *Solid State Ionics.* Vol.176 , No.3-4, (January 2005), pp. 385-392.

Preparation of NiO Catalyst on FeCrAl Substrate Using Various Techniques at Higher Oxidation Process

Darwin Sebayang[1], Yanuandri Putrasari[1], Sulaiman Hasan[1],
Mohd Ashraf Othman[1] and Pudji Untoro[2]
[1]Universiti Tun Hussein Onn Malaysia,
[2]Badan Tenaga Nuklir Nasional
[1]Malaysia
[2]Indonesia

1. Introduction

Catalytic converter consists of three major components, i.e substrate, catalyst, and washcoat. The first one is a substrate, a support for catalyst material. The FeCrAl is generally considered as metallic substrates due to their advantage in the high-temperature corrosion resistance, including the strong adherence of oxide film on the surface of substrate when applied the appropriate surface treatment (Twigg & Webster, 2006; Pilone, 2009; Klower et al., 1998; Cueff et al., 2004; Badini & Laurella; 2001; Czyrska-Filemonowicz et al., 1999; Liu et al., 2001; Amano et al., 2008; Checmanowski & Szczygiel, 2008). This material is based on ferritic steels with 5-8 wt% aluminium, 17-22 wt% chromium, plus small additions of reactive elements, which are added to improve the oxidation resistance and oxide adhesion (Nicholls & Quadakkers, 2002). Meanwhile, a catalyst is the accelerate agent for chemical reaction in terms of oxidation and reduction of emission gas. The existing of excellent oxidation catalyst materials was usually based on the precious metal (Pt, Pd, and Rd). However, those materials are expensive and limited supply (Koltsakis & Stamatelos, 1997; Benson et al., 2000). For this reason, the cheaper ranges of oxides (e.g. CuO, V_2O_5, NiO, MoO_3, and Cr_2O_3) compared to precious metals are being investigated as an alternative catalyst (Kolaczkowski, 2006). This work reports the use of NiO catalyst developed from Ni as a starting material. A washcoat is a catalysts carrier with high surface area. This material is usually an oxide layer such as Al_2O_3, SiO_2, TiO_2, or SiO_2-Al_2O_3 (Heck et al., 2002). Nickel forms under a normal temperature and pressure conditions only one oxide, NiO. The mechanism by which oxidation of a nickel proceeds was involved the outward migration of cations and electrons, which forming a single-phase scale (Birks et al., 2006). The conventional technique for adhering catalyst on substrate is by washcoating techniques that generally comprise of preparing a coating formed from a high surface area oxide blended with one or more catalysts and dipping the monolith structure into that coating blend (Huang & Bar-Ilan, 2003; Eleta et al., 2009). One of the most common methods to form a thin layer of oxide coatings on the metallic substrate is dip coatings, which combined with some pre-treatments, such as growing a number of textured alumina whiskers on the surface of the metal support and shortened the diffusion path before depositing the washcoat (Zao, et al., 2003; Jia et al., 2007). The other methods are co-precipitation, sol-gel and spray-pyrolysis methods were also applied for preparation of FeCrAlloy supported perovskite for catalytic combustion of methane (Yanqing, et al., 2010). Furthermore, another technological procedure to develop and adhere to the catalysts on the FeCrAl substrate are based on electrophoretic deposition (Sun et al., 2007; Corni et al., 2008), solution combustion synthesis (SCS) (Specchia et al., 2004), aluminizing technique (Wu et al., 2007), and hydrothermal method (Zamaro et al., 2008; Wei et al., 2005; Mies et al., 2007; Sivaiah, 2010). However, the existing methods still have some limitations, especially due to the rather complicated method to applying the catalyst which is in the form of powder. This paper presents an innovative method for preparation of NiO catalyst on FeCrAl substrate through the combination of electroplating, ultrasonic treatment and oxidation process. Electroplating method was applied to coat Ni to the FeCrAl. The ultrasonic treatment was used in order to accelerate the solid particles to high velocities (Suslick et al., 1999). And, the oxidation process was aimed to convert Ni into NiO on the FeCrAl surface and to develop Al_2O_3 layer as well.

2. Methodology

2.1 Materials

The FeCrAl foils strip (Aluchrom YHf) was supplied by ThyssenKrupp VDM, Germany. The chemical compositions of the specimen according to ThyssenKrupp Data Sheet No. 4049 are listed in Table 1. The Al_2O_3 powders, SiC powders, nickel sulphamate ($Ni(SO_3NH_2)_2.4H_2O$), nickel chloride ($NiCl\,6H_2O$), boric acid (H_2BO_3), sodium lauryal sulphate ($C_{12}H_{25}SO_4Na$), hydrochloric acid (HCl), sodium hydroxide (NaOH), methanol (99%), ethanol (99%) and nickel plates (high-purity Ni) were obtained from Sigma Aldrich, Sdn. Bhd. (Malaysia).

	Ni	Cr	Fe	C	Mn	Si	Al	Zr	Y	Hf	N
min	-	19.0	bal	-	-	-	5.5	-	-	-	-
max	0.3	22.0		0.05	0.50	0.50	6.5	0.07	0.10	0.10	0.01

Table 1. Chemical composition (wt. %) of FeCrAl

2.2 Experimental procedures

The approach started with assessment of FeCrAl treated by using various nickels electroplating process based on the weight gain during oxidation, followed by short term oxidation process, and long term oxidation process. The steps are summarized in flowchart as shown in Figure 1.

2.2.1 First step: Assessment of FeCrAl treated using various nickel electroplating process

In this study, the various electroplating processes of nickel on the FeCrAl were carried out as the preliminary study to obtain the optimum method to develop nickel oxide on the FeCrAl substrate.

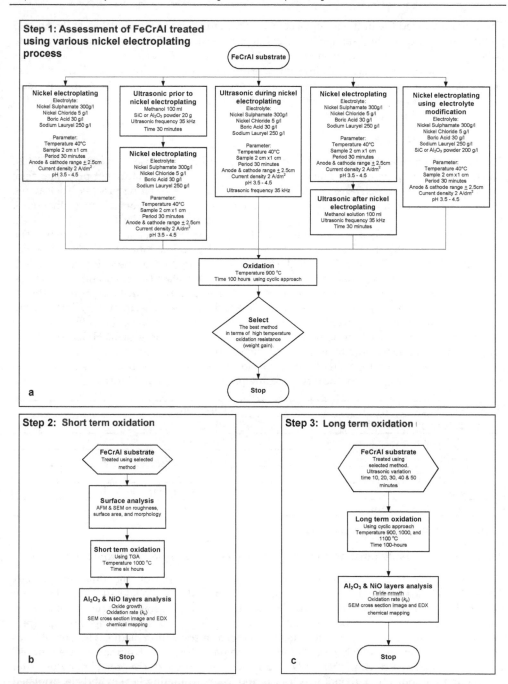

Fig. 1. Flow chart of the research, a) assessment of FeCrAl treated using various nickel electroplating, b) short term oxidation, c) long term oxidation

The study analyzed the influence of various electroplating processes of nickel on the FeCrAl metallic monolith for high-temperature oxidation resistance. The proposed new ideas to adhere to nickel as a catalyst on the FeCrAl substrate is divided into five methods as follows: Nickel electroplating, ultrasonic treatment prior to, during, and after nickel electroplating, and nickel electroplating using electrolyte modification. The optimum result in terms of high-temperature oxidation resistance which obtained from this investigation was then selected for further study/testing.

2.2.1.1 Nickel electroplating

The nickel electroplating process was conducted according to Rose & Whittington (2002). The following equipments, e.g. laboratory power supply, retorch clamp, hot plate magnetic steering, glass beaker, petri disc, and, etc. were used for electroplating process. The sample testing of FeCrAl foil was cut into 2 cm x 1 cm, and surface cleaned using ethanol. A solution was prepared by mixing 300 g/l of nickel sulphate powder and 5 g/l of nickel chloride powder, which dissolved into one litre of distilled water in a beaker glass. The concentration of 30 g/l of boric acid was then added to the solution, and pH value was maintained between 3.5 - 4.5. Boric acid acts as a buffer, to control the pH of the solution. The FeCrAl substrate was attached as a cathode and nickel plate (4 cm x 1 cm) as an anode. Both specimens then dipped simultaneously into the solution. The distance of the cathode-anode was set at the minimum 2.5 cm. The current density was setup at 2 A/dm². During the experiment, bubbles occurred at the surface of the sample and the pH was fluctuated. To remove the bubbles, 0.1% of sodium lauryl sulphate (SLS) was added. H_2SO_4 or NaOH solutions were added to maintain the pH. The electroplating process was started after 30 minutes. The specimen then dried. The condition of electroplating process is presented in Table 2, and the illustration of the electroplating process is shown in Figure 2.

Electroplating Condition	Parameter
Temperature of electroplating	40 °C
pH of electrolyte	3.5-4.5
Size of sample	2 cm x 1 cm
Electroplating period	30 minutes
Type of bath	Nickel sulphamate bath
Anode and cathode range	Min. 2.5 cm

Table 2. Electroplating process condition

2.2.1.2 Agitation using ultrasonic prior to, during, and after nickel electroplating

The *Fritsch Loborette 17* ultrasonic cleaning bath was used to conduct the ultrasonic process. The technical data in the ultrasonic apparatus were as follows: Voltage of 230 V/1~, input power of 2 x 240 W/period, frequency of 50-60 Hz and the ultrasound frequency of 35 kHz. Figure 3 (i) shows the condition for ultrasonic prior to electroplating process (pre-treatment). In this method, not only using methanol as a sonication media, Al_2O_3 or SiC was also mixed with methanol 200 g/l. Meanwhile, the particle size distribution of Al_2O_3 or SiC powders was analyzed using *Cilas 1180*. The sonication was conducted for 30 minutes. After completing the sonication, then it was transferred to electroplating beaker and electroplated with similar to the normal electroplating process. The sonication process of the sample was carried out by mixing the methanol with Al_2O_3 or SiC, which called as pre-treatment.

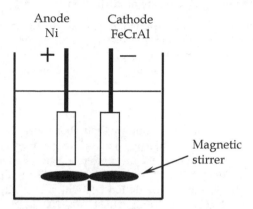

Fig. 2. Electroplating of FeCrAl alloy with nickel in a nickel sulphamate bath

Instead of just electroplating, ultrasonic process is also used to assist the nickel electroplating process. The condition of the electroplating process was similar with the previous, except the magnetic stirrer was replaced by ultrasound. The schematic diagram of the process is shown in Figure 3 (ii).

The ultrasonic process was also utilised after nickel electroplating process. In this method, the condition is similar to normal electroplating except the sonification was carried out after the electroplating process completed. The specimen was added into beaker glass with methanol solutions then sonicated for 30 minutes. The schematic diagram of the process is shown in Figure 3 (i).

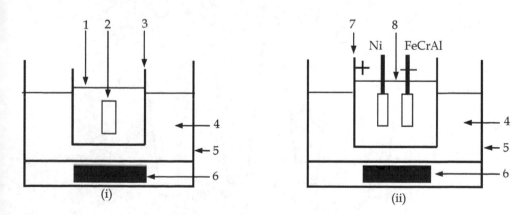

Fig. 3. (i) Schematic diagram of ultrasonic prior to (pre-treatment) and after electroplating process; (ii) Schematic diagram ultrasonic during electroplating process. (1) methanol; (2) specimen; (3) beaker; (4) water; (5) bath; (6) ultrasonic source; (7) plating tank; (8) electrolyte solution

2.2.1.3 Nickel electroplating using electrolyte modification

The new concept of electroplating was applied in this study to develop washcoat onto FeCrAl substrate, which done by mixing the electrolyte with 200 g/l of Al_2O_3 or SiC powder. This method is similar to normal electroplating except the electrolyte was modified. During electroplating process, the electrolyte was agitated using a magnetic stirrer to dissolve Al_2O_3 or SiC powder.

2.2.1.4 Oxidation process

The samples which produced by each variation method above were then oxidized to form the certain oxide (Al_2O_3 or NiO). The uncoated FeCrAl substrate was also oxidized. The oxidation test in this study was conducted according to previous work by Badini & Laurella (2001) and ASTM G 54-84 (1996) standard. The *Carbolite* automatic high-temperature furnace model HTF 1800 was used for isothermal oxidation test with a cyclic approach (Nicholls & Quadakkers, 2002; Fukuda, et al., 2002; Taniguchi, et al., 2002; Lylykangas & Tuomola, 2002). The test was carried out for 100 hours. The isothermal oxidation test with a cyclic approach is illustrated in Figure 4. The specimens were prepared by cutting them carefully into 5 mm x 5 mm. Then, the specimens were put in 5 mm diameter alumina crucible bucket. The weight of both specimen and bucket had been determined and recorded prior to oxidation test. The specimen that put in the alumina crucible bucket then delivered to the automatic furnace. The temperature was set-up of 30 to 910 °C of maximum temperature due to the catalytic converter working condition (Heck et al., 2002). The heating and cooling rate was set in 5 °C/minutes and holding time of 20 hours for each cycle. Then, the weight changes in every 20 hours were recorded. The testing was carried out until five cycle oxidation process. The weight of a specimen was measured, and the data was recorded after each cycle finished. The weight gains versus time graph were then plotted.

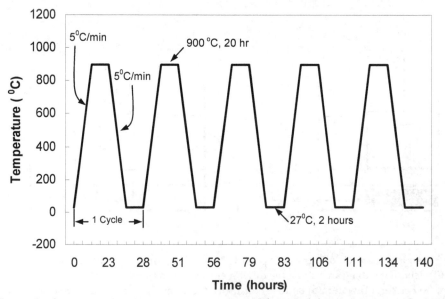

Fig. 4. Cyclic approach testing program (Adopted from Lylykangas & Tuomola, 2002)

2.2.2 Second step: Short term oxidation process

Prior to short term oxidation, the surface FeCrAl substrate was analysed in order to obtain the information of substrate surface, which previously treated using the selected method. The surface analysis was carried out by using step-and-scan automation of Atomic Force Microscopy (AFM) by *Park* model XE-100. Roughness (Ra) measurement was taken by horizontal straight-line mode. The Ra profiles were analysed and presented in order to clarify the roughness which caused by ultrasonic treatment of the FeCrAl surface specimens. The mean Ra is a result from a random 100 μm^2 scan area of specimens. The 3D images from AFM were presented to analyze the topography of a specimen. The grain area (μm^2) as 2D image was measured by using AFM analysis to calculate the total surface area of a specimen. The scanned surface areas of a specimen were calculated and approached by spherical surface analysis (Henke et al., 2002). This approach was adopted for this research. The approach assumed that the grain morphologies as a nodule on FeCrAl substrate in half of sphere form. Then, the mean area obtained from AFM was assumed as an area of circles where the diameter similar to the sphere. The half of sphere surface area was calculated by using sphere surface area formula. The surface analyses were carried out after ultrasonic pre-treatment without nickel electroplating to observe any changes in surface characteristic. The short term oxidation analysis focused on weight gain, oxidation rate (k_p), morphology of oxide layer and cross section elemental mapping of washcoat (Al_2O_3 layer). Meanwhile, the oxidation test for six hours was conducted using *Diamond, Perkin Elmer* thermo gravimetric analyzer (TGA). The specimens were prepared by cutting into 2 mm x 2 mm. All of specimens are tested for oxidation by putting them into the TGA at 1000 °C for 360 minutes. Then, the data plotted into the graph of weight gain (mg/mm^2) versus temperature (°C) or time (minutes). The graph was analyzed to obtain the parabolic rate constant (Smallman & Bishop, 1999; Badini & Laurella, 2001). The parabolic growth equation of the film thickness with time obtained by:

$$x^2 = k_p \cdot t \tag{1}$$

where x is the layer thickness or the weight gain; t is the oxidation treatment time; k_p is the parabolic rate constant. In this study, the weight gain (x) is a resulted from mass gain per unit surface area of specimens $\left(\dfrac{\Delta W}{A}\right)$ (Badini & Laurella, 2001). Then, the eq. (1) can be written as follow:

$$\left(\frac{\Delta W}{A}\right)^2 = k_p \cdot t \tag{2}$$

where k_p is obtained from the slope of a linear regression-fitted line of $\left(\dfrac{\Delta W}{A}\right)^2$ vs t plot.

The microstructure analyses were carried out using JEOL Scanning Electron Microscope (SEM) model JSM-6380LA attached with Energy Dispersive X-ray (EDX). Prior to microstructure analysis, the specimen is mounted, ground and polished and coated with gold or platinum. The sample mounting conducted by hot press and cold mounting technique at a cross-sectional side of the specimen. The *Buehler* automatic hot mounting press machine was implemented for hot mounting process. The mounting parameters were 15 minutes mounting period, 2000 psi mounting pressure and 150 °C as the mounting temperature. For the cold mounting process, the ratio composition of resin and hardener was 10 : 1. Then, the specimens were put into a mould at room temperature for minimum 24 hours until hardened. The specimens were ground using the SiC paper from 240 to 2000 grit, followed by polishing process, in order to obtain clear and shiny surface specimens.

After polishing, the specimens were observed under the digital microscope to ensure that the specimen did not have any scratches before further analyzed by SEM/EDX. The specimen was then coated with gold or platinum by using the sputter coating apparatus. The sputter coating was set in 20 mA of coating current and 20 minutes of coating time.

The back scattered mode of SEM was used to obtain a high-quality image of the specimen. The ranges of magnification, 100 x, 500 x, 1000 x, 1500 x and 2000 x, were used to observe both surface and cross-section of the specimen. The EDX mapping and line analysis techniques were also implemented in order to reveal the distribution of oxide layers on the FeCrAl surface substrate.

2.2.3 Third step: Long term oxidation process

This step is aimed to explore the behaviour of nickel layer on FeCrAl when subjected to the selected treatment at variation time for 10, 20, 30, 40 and 50 minutes, with high oxidation temperature for 100 hours. The oxidation process conducted using a cyclic approach testing, similar with previous sub section 2.2.1.4. To assess the high-temperature oxidation resistance of FeCrAl, the temperature oxidation of 1000 and 1100 °C were also applied. As similar with short term oxidation, in this step the oxide growth, and oxidation rate (k_p) were also analysed. Nickel and nickel oxide morphologies were studied on the samples cross section at various temperatures. EDX attached to SEM was also used to obtain the elemental distribution of the samples cross section similar with sub section 2.2.2.

3. Results and discussion

3.1 Assessment of FeCrAl treated using various nickel electroplating process based on weight gain

The results of the isothermal oxidation at 900 °C in the air by cyclic oxidation test approach using an automatic high-temperature furnace presented in the graph of weight gain versus time as shown in Figure 5. The high weight gain of the FeCrAl due to poor high-temperature resistance. With more oxides developed during the elevated temperature, it will cause the substrate thickness of FeCrAl to decrease due to the transformation of its component to oxide. From this experiment, it can be seen that the electroplating of nickel will be a good barrier for the high-temperature resistance if treated with ultrasonic as well. The adhesions of nickel deposition on FeCrAl surface was improved by using the ultrasonic process. In this case, the ultrasonic process cause the collapse of cavitations bubbles of liquid and generates a pressure shock wave, liquid jet, and water hammer pressure. The ultrasonic act as an agitator, make faster stirring process. Thus, the electrolyte will mix properly as it is needed to fulfil the requirement of electroplating process (Chiba et al., 2000). From the results as mentioned above, it can be concluded that the ultrasonic pre-treatment cause the oxide grow sufficiently on the FeCrAl substrate and expected to have the capability for high-temperature oxidation resistance. Therefore, this method is selected and applied for further study.

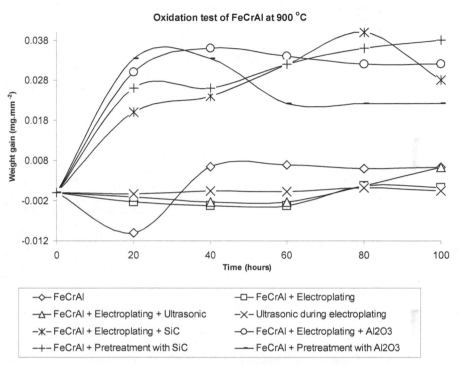

Fig. 5. Effect of various electroplating process on weight gain vs time of FeCrAl during oxidation at 900 ºC

3.2 Analysis of FeCrAl substrate before and after short term oxidation

3.2.1 Surface analysis

Figure 6 shows the roughness profile of the FeCrAl (a) untreated, (b) ultrasonic treatment with SiC for 10 minutes and (c) ultrasonic treatment with Al_2O_3 for 10 minutes without nickel electroplating. From these profiles, the highest, medium, and lowest gap between peaks and valley occurred on FeCrAl untreated, treatment with Al_2O_3 and with SiC, respectively.

The mean roughness of FeCrAl was presented in Table 3. The mean roughness of each specimen was obtained by using the horizontal straight-line method on random position of 10 µm x 10 µm image. Meanwhile, the 3D images for all specimens were presented in Figure 7.

Materials	Mean Roughness, Ra (nm)
FeCrAl untreated	31.409
FeCrAl ultrasonic treatment with SiC	15.790
FeCrAl ultrasonic treatment with Al_2O_3	34.470

Table 3. Mean roughness of FeCrAl surface

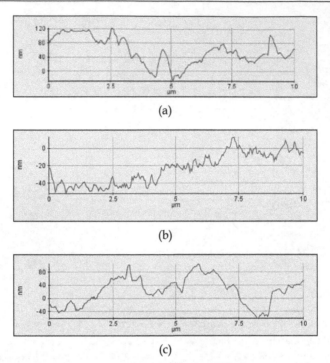

(a)

(b)

(c)

Fig. 6. Roughness profile of FeCrAl surface a) untreated, b) ultrasonic treatment with SiC for 10 minutes, c) ultrasonic treatment with Al_2O_3 for 10 minutes

Based on the roughness test results, the FeCrAl treated with Al_2O_3 has the highest surface roughness value compared to the untreated and treated using ultrasonic with SiC. Based on the particle size distribution analysis that was conducted, for 60% distribution of Al_2O_3 powder is 42.07 μm with specific surface 5658.19 cm^2/g and 60% distribution of SiC powder is 87.56 μm with specific surface only 2279.44 cm^2/g. It can be estimated that the roughness of FeCrAl surface depends on the particle size and homogeneity of the powders. The surface roughness of FeCrAl was as also estimated can be increased by the higher specific area of powders.

Figure 8 shows the grains of FeCrAl (a) untreated, (b) ultrasonic treatment with SiC for 10 minutes and (c) ultrasonic treatment Al_2O_3 for 10 minutes. These images resulted from 10 μm x 10 μm random scanning area on each specimen. The grain numbers and grain area were presented in Table 4. The table shows that the grain numbers on FeCrAl ultrasonic treatment with SiC is the highest, followed by untreated FeCrAl, and the lowest is FeCrAl ultrasonic treatment with Al_2O_3. The highest grain area is FeCrAl ultrasonic treatment with Al_2O_3, then FeCrAl untreated, and the smallest is FeCrAl ultrasonic treatment with SiC.

Materials	Grain numbers	Mean area (μm^2)
FeCrAl untreated	172	5.412×10^{-1}
FeCrAl ultrasonic treatment with SiC	183	5.058×10^{-1}
FeCrAl ultrasonic treatment with Al_2O_3	167	5.599×10^{-1}

Table 4. Mean of grain area on FeCrAl surface

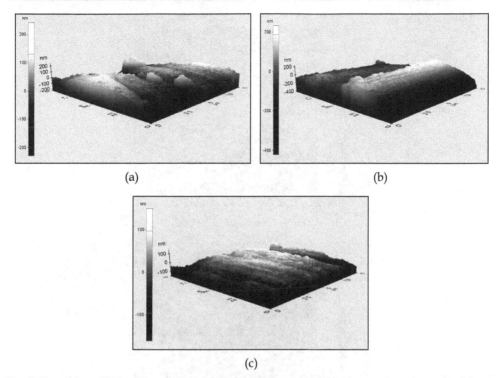

(a)   (b)

(c)

Fig. 7. Roughness 3D images of FeCrAl surface a) untreated, b) ultrasonic treatment with SiC for 10 minutes, c) ultrasonic treatment with Al_2O_3 for 10 minutes

The surface area acts as a main role on the catalyst reaction effectiveness. In order to accommodate the catalyst in significant amounts, substrate must be provided with a high surface area. Twigg & Webster (2006) suggest that the design of substrate must provide a maximum superficial surface area which accommodates to the exhaust gas, as it is upon this surface that the catalytic coating is applied, and on which the pollutant and reactant gases must impinge in order to react.

3.2.2 Parabolic rate constant

The parabolic rate constant (k_p) can be used to predict the time to failure of the FeCrAl materials (Klower et al., 1998). The formation rate of an oxide scale, growing on the surface of a FeCrAl surface at the beginning of the oxidation test agrees with the Wagner theory. At high-temperature oxide, films are thickened according to the parabolic rate law, $x^2 \propto t$ and the mechanism by which thickening proceeded has been explained by Wagner (Smallman & Bishop, 1999; Badini & Laurella, 2001).

Figure 9 shows the nature of the fit of the parabolic rate law of the early oxidation test of FeCrAl substrate pre-treatment using ultrasonic with SiC or Al_2O_3 at 1000 °C for 60 minutes. The parabolic rate constants obtained from the present experiments are listed Table 5. FeCrAl substrate pre-treated with SiC has lower k_p than the FeCrAl substrate pre-treated

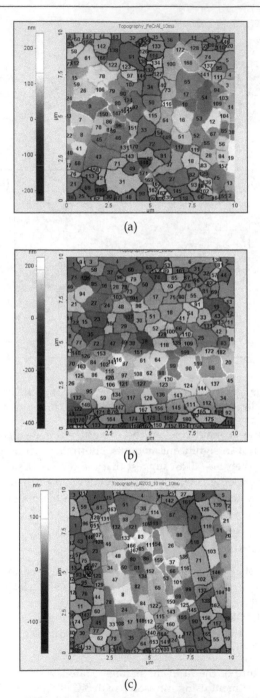

Fig. 8. Grain area image of FeCrAl surface a) untreated, b) ultrasonic treatment with SiC for 10 minutes, c) ultrasonic treatment with Al₂O₃ for 10 minutes

with Al_2O_3. The low of k_p value indicated the long time to failure of FeCrAl substrate (Klower et al., 1998). Thus, from these finding the FeCrAl substrate treated with SiC has better time to failure than treated with Al_2O_3.

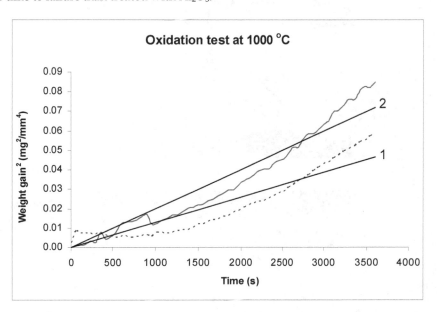

Fig. 9. $\left(\dfrac{\Delta W}{A}\right)^2$ vs time plotted for oxidation of FeCrAl pre-treatment using ultrasonic prior to nickel electroplating with SiC (1) and with Al_2O_3 (2)

Materials	$k_p \times 10^{-6}$ mg^{2}mm^{-4}s^{-1}
FeCrAl substrate pre-treated with SiC	13
FeCrAl substrate pre-treated with Al_2O_3	20

Table 5. Parabolic rate constants for FeCrAl substrate pre-treated with SiC or Al_2O_3 at 1000°C

3.2.3 Cross section analysis of Al_2O_3 and NiO layers

Figure 10 shows the cross section scanning micrograph of the FeCrAl substrate pre-treatment using the ultrasonic process with SiC and Al_2O_3 powder prior to nickel electroplating after short term oxidation. The cross-section of the images shows four layers. The first layer from bottom to top is FeCrAl substrate followed by Al_2O_3 as the second layer, and the third is the nickel layer, and the fourth layer is the nickel oxide layer. The cross section observation showed that the NiO layer existed on the FeCrAl ultrasonic with SiC or Al_2O_3 after short term oxidation.

Based on surface analysis, the ultrasonic process increased irregular surface roughness morphology on FeCrAl substrate. It might influence the homogeneous and stability of nickel electroplating, and also for NiO surface area development. The nickel electroplating on FeCrAl ultrasonic with SiC before electroplating showed more homogeneous, and it is a

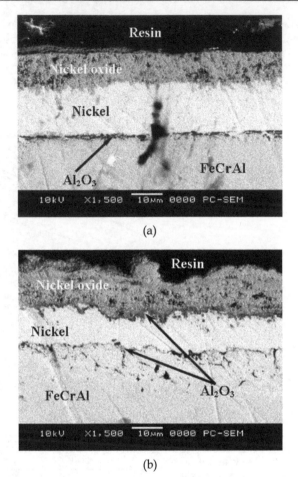

Fig. 10. Cross-section scanning electron micrograph showing four layers of FeCrAl pre-treatment using ultrasonic with (a) SiC and (b) Al_2O_3 prior to nickel electroplating after short term oxidation

more stable condition than Al_2O_3. Thus better density of nickel layer was achieved. Possibly, this was influenced by high surface roughness resulted from the FeCrAl substrate pre-treatment using ultrasonic with SiC. The cross section image clearly indicated that Al_2O_3 and NiO layers occurred due to the heating process. However, the nickel layer still existed after short oxidation, so that needs further study.

3.3 Analysis of Al_2O_3 and NiO layers for long term oxidation

3.3.1 Influence of various pre-treatment times and temperatures on weight gain and parabolic rate constant

The influence of various pre-treatment times on the weight gain of FeCrAl ultrasonic pre-treatment with SiC before electroplating and oxidized at 900, 1000, 1100 ºC are shown in

Figure 11, 12, and 13, respectively. The influence of various pre-treatment times on the weight gain of FeCrAl ultrasonic pre-treatment with Al_2O_3 before electroplating after oxidation at 900, 1000, and 1100 ºC are presented also as a graph in Figure 14, 15, and 16. The increasing of the graph shows the growth of the oxide layer. Meanwhile, the decreasing graph indicated the oxide layer spallation. All the specimens with different ultrasonic treatment and different temperature oxidation represent the growth of the oxide layer. The specimen without spallation at 900 ºC is the FeCrAl ultrasonic pre-treatment with SiC for 10 minutes only. At 1000 ºC, FeCrAl pre-treatment with SiC 10 minutes, Al_2O_3 10 minutes, and Al_2O_3 30 minutes show no spallation. For the oxidation at 1100 ºC, the FeCrAl pre-treatment with SiC 10 minutes and Al_2O_3 20 minutes also shows no spallation.

Both types of FeCrAl ultrasonic pre-treatment graphs were analyzed to obtain the estimation of parabolic rate constant as listed in Table 6 for FeCrAl ultrasonic pre-treatments with SiC or Al_2O_3 oxidized at 900 ºC, Table 7 for 1000 ºC and Table 8 for 1100 ºC. The lowest parabolic rate constants of FeCrAl ultrasonic pre-treatment with SiC are shown by its sonication time, as follows: after oxidation at 900 ºC for 10 minutes, 1000 ºC for 10 minutes, and 1100 ºC for 30 minutes. Meanwhile, the lowest parabolic rate constant of FeCrAl ultrasonic pre-treatments with Al_2O_3 are oxidized at 900 ºC for 30 minutes, 1000 ºC for 30 minutes, and 1100 ºC for 30 minutes.

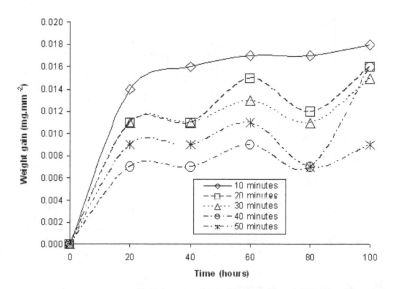

Fig. 11. Influence of various pre-treatment times on weight gain of FeCrAl treated using ultrasonic with SiC prior to nickel electroplating during oxidation at 900 ºC using cyclic approach

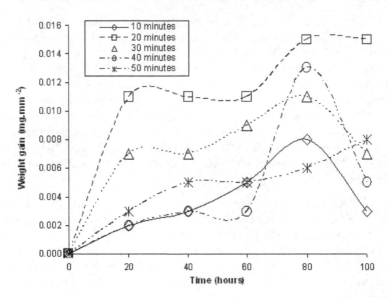

Fig. 12. Influence of various pre-treatment times on weight gain of FeCrAl treated using ultrasonic with SiC prior to nickel electroplating during oxidation at 1000 ºC using cyclic approach

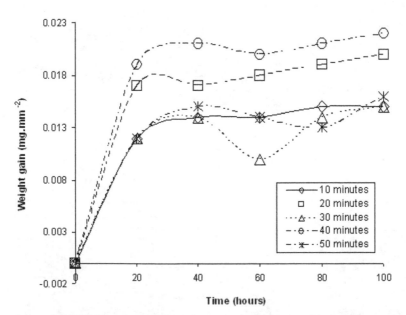

Fig. 13. Influence of various pre-treatment times on weight gain of FeCrAl treated using ultrasonic with SiC prior to nickel electroplating during oxidation at 1100 ºC using cyclic approach

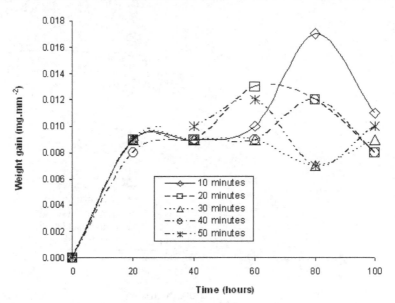

Fig. 14. Influence of various pre-treatment times on weight gain of FeCrAl treated using ultrasonic with Al_2O_3 prior to nickel electroplating during oxidation at 900 °C using cyclic approach.

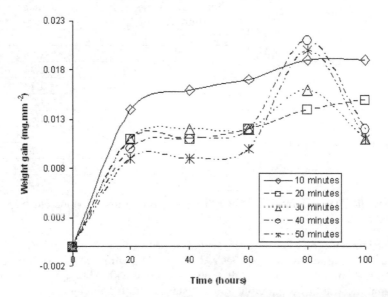

Fig. 15. Influence of various pre-treatment times on weight gain of FeCrAl treated using ultrasonic with Al_2O_3 prior to nickel electroplating during oxidation at 1000 °C using cyclic approach

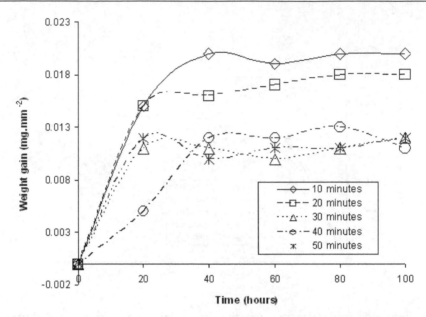

Fig. 16. Influence of various pre-treatment times on weight gain of FeCrAl treated using ultrasonic with Al_2O_3 prior to nickel electroplating during oxidation at 1100 °C using cyclic approach

Specimens	Sonication Time (minutes)	Parabolic Rate Constant k_p x 10^{-4} $mg^2mm^{-4}h^{-1}$
FeCrAl ultrasonic pre-treatment with SiC	10	2.32
	20	1.87
	30	1.74
	40	1.42
	50	1.21
FeCrAl ultrasonic pre-treatment with Al_2O_3	10	1.64
	20	1.40
	30	1.15
	40	1.28
	50	1.30

Table 6. Parabolic rate constant (k_p) of FeCrAl treated using ultrasonic and electroplating methods and oxidized at 900 °C

The table of parabolic rate constant of oxidation test at 900 °C showed that the lowest parabolic rate constant obtained from FeCrAl ultrasonic treatment with SiC for 50 minutes at 1.21 x 10^{-4} $mg^2mm^{-4}h^{-1}$, and FeCrAl ultrasonic treatment with Al_2O_3 for 30 minutes at 1.15 x 10^{-4} $mg^2mm^{-4}h^{-1}$. The lower parabolic rate constant indicated the longer time to failure of the FeCrAl substrate (Klower, et al., 1998). It can be seen from the table of parabolic rate constant at 900 °C, that longer pre-treatment process with SiC influenced to lower parabolic constant of FeCrAl, but it was not fully applied in pre-treatment with Al_2O_3.

Specimens	Sonication Time (minutes)	Parabolic Rate Constant k_p x 10^{-4} mg^{2}mm^{-4}h^{-1}
FeCrAl ultrasonic pre-treatment with SiC	10	0.64
	20	1.83
	30	1.15
	40	0.85
	50	0.84
FeCrAl ultrasonic pre-treatment with Al$_2$O$_3$	10	2.44
	20	1.82
	30	1.73
	40	1.92
	50	1.75

Table 7. Parabolic rate constant (k_p) of FeCrAl treated using ultrasonic and electroplating methods and oxidized at 1000 °C

Specimens	Sonication Time (minutes)	Parabolic Rate Constant k_p x 10^{-4} mg^{2}mm^{-4}h^{-1}
FeCrAl ultrasonic pre-treatment with SiC	10	1.97
	20	2.55
	30	1.83
	40	2.86
	50	1.96
FeCrAl ultrasonic pre-treatment with Al$_2$O$_3$	10	2.65
	20	2.36
	30	1.52
	40	1.56
	50	1.54

Table 8. Parabolic rate constant (k_p) of FeCrAl treated using ultrasonic and electroplating methods and oxidized at 1100 °C

In FeCrAl after oxidation at 1000 °C, the lowest parabolic rate constant for pre-treatment with SiC obtained for 10 minutes and with Al$_2$O$_3$ for 30 minutes. There was no linear relation between ultrasonic pre-treatment and parabolic rate constant oxidation test at 1000 °C. However, there are some results by using ultrasonic with SiC powders obtained lower parabolic rate constant, under 1 x 10^{-4} mg^{2}mm^{-4}h^{-1}, than pre-treatment with Al$_2$O$_3$ after oxidation at 1000 °C. The very low parabolic rate constant of FeCrAl pre-treatment ultrasonic with SiC occurred for 10 minutes in 0.64 x 10^{-4} mg^{2}mm^{-4}h^{-1}, for 50 minutes in 0.84 x 10^{-4} mg^{2}mm^{-4}h^{-1}, and for 40 minutes in 0.85 x 10^{-4} mg^{2}mm^{-4}h^{-1}. Meanwhile, ultrasonic pre-treatment with Al$_2$O$_3$ resulted in 1 x 10^{-4} mg^{2}mm^{-4}h^{-1} after oxidation at 1000 °C for 100 hours.

According to the parabolic rate constant table of specimens after oxidation at 1100 °C all the parabolic rates constant are higher than 1 x 10^{-4} mg^{2}mm^{-4}h^{-1}. From the parabolic rate constant results, it can be estimated that both kind FeCrAl ultrasonic pre-treatment with SiC or Al$_2$O$_3$ when oxidized at 1100 °C will be fail faster than those oxidized at 900 and 1100 °C. It is generally known that higher temperature caused faster material failures.

3.3.2 Cross section analysis of Al₂O₃ and NiO layers

Figure 17 presents cross section scanning electron micrograph with EDX line analysis of the NiO catalyst samples prepared on FeCrAl substrate through ultrasonic with SiC for 50 minutes (Figure 17.a), Al₂O₃ 30 minutes (Figure 17.b), then combined with nickel electroplating, and oxidation process at 900 ℃. The EDX line analysis shows that nickel layer disappeared and fully converted to nickel oxide. Until the end of the oxidation exposure, the nickel oxide still existed.

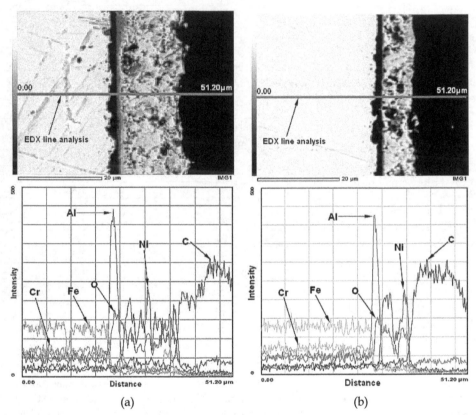

(a)   (b)

Fig. 17. Cross section scanning electron micrograph showing EDX line analysis (from left to the right) of FeCrAl ultrasonic with (a) SiC for 50 minutes, (b) Al₂O₃ for 30 minutes prior to nickel electroplating and oxidized at 900 ℃ with its chemical's intensity graph

Figure 18 shows cross section scanning electron micrograph with EDX line analysis of the NiO catalyst samples which developed on FeCrAl substrate through ultrasonic with SiC for 10 minutes (Figure 18.a), Al₂O₃ for 30 minutes (Figure 18.b), then combined with nickel electroplating, and oxidation process at 1000 ℃. According to the EDX line analysis, it is clear that both types of the FeCrAl whether treated by ultrasonic with SiC or Al₂O₃ then electroplated with nickel showed the phase change of nickel to nickel oxide and existing of nickel oxide during oxidation exposure for 100 hours.

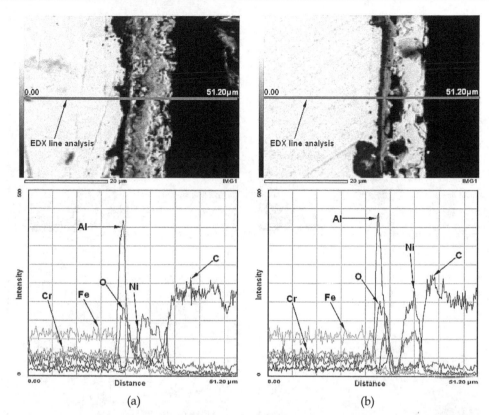

(a)   (b)

Fig. 18. Cross section scanning electron micrograph showing EDX line analysis (from left to the right) of FeCrAl ultrasonic with (a) SiC for 10 minutes, (b) Al_2O_3 for 30 minutes prior to nickel electroplating and oxidized at 1000 °C with its chemical's intensity graph

Figure 19 shows cross section scanning electron micrograph with EDX line analysis of the NiO catalyst samples which prepared on FeCrAl substrate through ultrasonic with SiC for 30 minutes (Figure 19.a), Al_2O_3 30 minutes (Figure 19.b), then combined with nickel electroplating, and oxidation process at 1100 °C. The nickel phase layer completely changed to be nickel oxide phase. It showed that the nickel oxide still existed during the oxidation exposure.

From the cross section SEM/EDX (Fig. 17, 18 and 19), it is clearly that the nickel layer disappeared due to full conversion to nickel oxide during 100 hour oxidation processes. The large amount of the nickel oxide present in the outer oxide layer. The nickel oxide still existed although it seems several spallations after the oxidation process. The spallation occurred due to the influence of 100 hours oxidation. Besides NiO and Al_2O_3 layers, according to chemical's graph there are several oxide layers or scale, which occurred as Cr_2O_3 and Fe_2O_3. The Al_2O_3 layer also occurred in stable condition as shown in the cross section image. Both specimens showed the evolution of nickel layer on the FeCrAl surface substrate, where it was fully converted to nickel oxide after oxidation exposure for 100 hours.

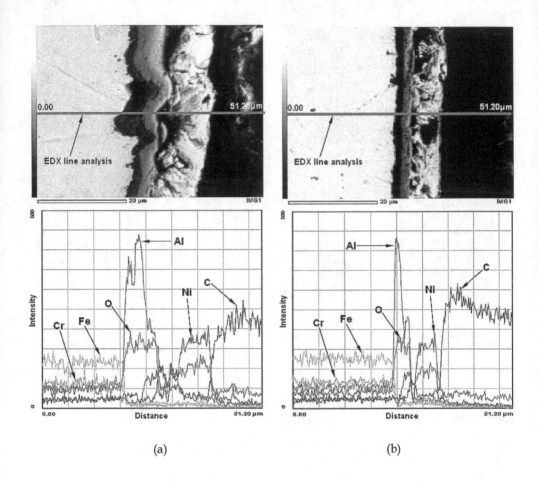

Fig. 19. Cross section scanning electron micrograph showing EDX line analysis (from left to the right) of FeCrAl ultrasonic with (a) SiC for 30 minutes, (b) Al₂O₃ for 30 minutes prior to nickel electroplating, and oxidized at 1100 ᵒC with its chemical's intensity graph

4. Conclusion

A systematic study was conducted to evaluate the new method of developing NiO through oxidation in combination between nickel electroplating and ultrasonic technique. The NiO was obtained when treated at short term of oxidation for 6 hours at 900 ᵒC and fully developed for 100 hours after oxidation at 900, 1000 and 1100 ᵒC. The NiO catalyst successfully developed from nickel plating, which obtained on the surface of FeCrAl substrate through the electroplating combined with the ultrasonic process and oxidation process. For future work, the chemical properties of nickel oxide catalyst prepared on

FeCrAl substrate through the ultrasonic technique combined with the electroplating and oxidation process will be further investigated.

5. Acknowledgements

The authors would like to thank the Ministry of Higher Education Malaysia and Universiti Tun Hussein Onn Malaysia (UTHM) through the funding support of Fundamental Research Grant Scheme (FRGS), Vot no. 0265 and 0361. The authors would also like to thank Dr. W.J. Quadakkers from Juelich Research Centre and the ThyssenKrupp VDM GmbH for providing the material.

6. References

Amano, T.; Takezawa, Y.; Shiino, A. & Shishido, T. (2008). Surface Morphology of Scale on FeCrAl (Pd, Pt, Y) alloys. *Journal of Alloys and Compounds*, Vol. 452, No. 1, pp. 16-22

American Society for Testing and Materilas [ASTM]. (1996). Standard Practice for Simple Static Oxidation Testing. America: G 54-84.

Badini, C. & Laurella, F. (2001). Oxidation of FeCrAl Alloy: Influence of Temperature and Atmosphere on Scale Growth Rate and Mechanism. *Surface and Coating Technology*, Vol. 135, No. 2-3, pp. 291-298

Benson, M.; Bennett, C.R.; Harry, J.E.; Patel, M.K. & M. Cross. (2000). The Recovery Mechanism of Platinum Group Metals from Catalytic Converters in Spent Automotive Exhaust Systems. *Resources, Conservation and Recycling*, Vol. 31, No. 1, pp. 1-7

Birks, N.; Meier, G.H. & F.S. Pettit. (2006). *Introduction to The High Temperature Oxidation of Metals, 2nd ed*, Cambridge University Press, ISBN 978-0-511-16089-9, New York, USA

Checmanowski, J.G. & Szczygiel, B. (2008). High Temperature Oxidation Resistance of FeCrAl Alloys Covered with Ceramic SiO_2–Al_2O_3 Coatings Deposited by Sol–gel Method. *Corrosion Science*, Vol. 50, No. 12, pp. 3581-3589

Chiba. A.; Gotou. T.; Kobayashi, K. & Wu, W. (2000). Influence of sonication of nickel plating in a nickel sulfamate bath. *Metal Finishing*, Vol. 98, No. 9, pp. 66-69

Corni, I.; Ryan, M.P. & Boccaccini, A.R. (2008). Electrophoretic deposition: From traditional ceramics to nanotechnology. *Journal of the European Ceramic Society*, Vol. 28, No. 7, pp. 1353-1367

Cueff, R.; Buscail, H.; Caudron, E.; Riffard, F.; Issartel, C. & El Meski, S. (2004). Effect of Reactive Element Oxide Coating on the High Temperature Oxidation Behaviour of FeCrAl Alloys. *Applied Surface Science*, Vol. 229, No. 1-4, pp. 233-241

Czyrska-Filemonowicz, A.; Szot , K.; Wasilkowska , A.; Gil , A. & Quadakkers, W.J. (1999). Microscopy (AFM, TEM, SEM) Studies of Oxide Scale Formation on FeCrAl Based ODS Alloys. *Solid State Ionics*, Vol. 117, No. 1-2, pp. 13-27

Eleta, A.; Navarro, P.; Costa, L. & Montes, M. (2009). Deposition of zeolitic coatings onto Fecralloy microchannels: Washcoating vs. in situ growing. *Microporous and Mesoporous Materials,* Vol. 123, No. 1-3, pp. 113–122

Fukuda, K.; Takao, K.; Hoshi, T. & Furukumi, O. (2002). Improved High Temperature Oxidation Resistance of REM Added Fe-20%Cr-5%Al Alloy by Pre-Annealing Treatment. In: *Materials Aspects in Automotive Catalytic Converters,* Hans Bode (Ed.), 59-82, Wiley-VCH Verlag GmbH, ISBN 3-527-30491-6, Weinheim, Germany

Heck, R. M.; Farrauto, R. J. & Gulati, S. T. (2002). *Catalytic Air Pollution Control Commercial Technology 3rd ed,* John Wiley & Sons, ISBN 978 -0-470-27503-0, New Jersey

Henke, L.; Nagy, N. & Krull, U.J. (2002). An AFM determination of the effects on surface roughness caused by cleaning of fused silica and glass substrates in the process. *Biosensors and Bioelectronics,* Vol. 17, pp. 547-555

Huang, Y. & Bar-Ilan, A. (2003). *Method for washcoating a catalytic material onto a monolithic structure.* U. S. Patent 6759358

Jia, L.; Shen, M. & Wang, J. (2007). Preparation and characterization of dip-coated γ-alumina based ceramic materials on FeCrAl foils. *Surface & Coatings Technology,* Vol. 201, No. 16-17, pp. 7159–7165

Klöwer, J.; Kolb-Telieps, A.; Bode, H.; Brede, M.; Lange, J.; Brück, R. & Wieres, L. (1998). Development of high-temperature corrosion resistant FeCrAl alloys for automotive catalytic converters. *Materials Week Congress for Innovative Materials, Processes, and Applications,* Munich, Germany, 12-15.10.1998

Kolaczkowski, S. (2006). Treatment of Volatile Organic Carbon (VOC) Emissions from Stationary Sources: Catalytic Oxidation of The Gaseous Phase. In: *Structured Catalysts and Reactors 2nd ed,* A. Cybulski. & J. A. Moulijn, (Eds.), 147-169, Taylor & Francis Group, ISBN 0-8247-2343-0, Boca Raton, FL

Koltsakis, G.C. & Stamatelos, A.M. (1997). Catalytic Automotive Exhaust After Treatment. *Progress in Energy and Combustion Science,* Vol. 23, No. 1, pp. 1-39

Liu, G.; Rozniatowski, K. & Kurzydlowski, K.J. (2001). Quantitative Characteristics of FeCrAl Films Deposited by Arc and High-velocity Arc Spraying. *Materials Characterization,* Vol. 46, No. 2-3, pp. 99-104

Lylykangas,R. & Tuomola, H. (2002). A New Type of Metallic Substrate. In: *Materials Aspects in Automotive Catalytic Converters,* Hans Bode (Ed.), 152-170, Wiley-VCH Verlag GmbH, ISBN 3-527-30491-6, Weinheim, Germany

Mies, M.J.M.; Rebrov, E.V.; Jansen, J.C.; Croon, M.H.J.M. & Schouten, J.C. (2007). Hydrothermal synthesis of a continuous zeolite Beta layer by optimization of time, temperature and heating rate of the precursor mixture. *Microporous and Mesoporous Materials,* Vol. 106, No. 1-3, pp. 95–106

Nicholls, J. R. & Quadakkers, W. J. (2002). Materials Issues Relevant to the Development of Future Metal Foil Automotive Cataltic Converters. In: *Materials Aspects in Automotive Catalytic Converters,* Hans Bode (Ed.), 31-48, Wiley-VCH Verlag GmbH, ISBN 3-527-30491-6, Weinheim, Germany

Pilone, D. (2009). Ferritic Stainless Steels for High Temperature Applications in Oxidizing Environments. *Recent Patents on Materials Science*, Vol. 2, No. 1, pp. 27-31, ISSN 1874-4648

Rose, I. & Whittington, C. (2002*). Nickel Plating Handbook*. OMG Group, Finland

Sivaiah, M.V.; Petit, S.; Beaufort, M.F.; Eyidi, D.; Barrault, J.; Batiot-Dupeyrat, C. & Valange, S. (2010). Nickel based catalysts derived from hydrothermally synthesized 1:1 and 2:1 phyllosilicate as precursors for carbon dioxide reforming of methane. *Microporous and Mesoporous Materials*, Vol. 140, No. 1-3, pp. 69-80

Smallman, R.E. & Bishop, R.J. (1999). *Modern Physical Metallurgy and Materials Engineering Science, Process, Applications 6th ed.* Butterwort-Heinemann Elsevier, ISBN 0-7506-4564-4, Oxford.

Specchia, A.; Civera, A. & Saracco, G. (2004). In situcombustion synthesis of perovskite catalysts for efficient and clean methane premixed metal burner. *Chemical Engineering Science*, Vol. 59, No. 22-23, pp. 5091-5098

Sun, H.; Quan, X.; Chen,S.; Zhao, H. & Zhao, Y. (2007). Preparation of well-adhered g-Al$_2$O$_3$ washcoat on metallic wire mesh monoliths by electrophoretic deposition. *Applied Surface Scienc*, 253, pp. 3303–3310

Suslick K.S.; Didenko, Y.; Fang, M.; Hyeon, T.; Kolbeck, K.J.; Mc Namara III, W.B.; Mdleleni, M.M. & Wong, M. (1999). Acoustic cavitation and its chemical consequences. *Philosophical Transactions: Mathematical, Physical and Engineering Sciences*, Vol. 357, pp. 335-353

Taniguchi, S., Andoh, A., & Shibata, T. (2002). Improvement in The Oxidation Resistance of Al-deposited Fe-Cr-Al Foil by Pre-oxidation. In: *Materials Aspects in Automotive Catalytic Converters*, Hans Bode (Ed.), 83-105, Wiley-VCH Verlag GmbH, ISBN 3-527-30491-6, Weinheim, Germany

Twigg, M. V. & Webster, D. E. (2006). Metal and Coated Metal Catalysts. In: *Structured Catalysts and Reactors 2nd ed*, A. Cybulski. & J. A. Moulijn, (Eds.), 71-108, Taylor & Francis Group, ISBN 0-8247-2343-0, Boca Raton, FL

Wei, Q.; Chen, Z.X.; Nie, Z.R.; Hao, Y.L.; Zou, J.X. & Wang, Z.H. (2005). Mesoporous activated alumina layers deposited on FeCrAl metallic substrates by an in situ hydrothermal method. *Journal of Alloys and Compounds*, Vol. 396, No. 1-2, pp. 283–287

Wu, X.; Weng, D.; Zhao, S. & Chen, W. (2007). Influence of an aluminized intermediate layer on the adhesion of a g-Al$_2$O$_3$ washcoat on FeCrAl. *Surface & Coatings Technology*, Vol. 190, No. 2-3, pp. 434– 439

Yanqing, Z.; Jieming, X.; Cuiqing, L.; Xin, X. & Guohua, L. (2010). Influence of preparation method on performance of a metal supported perovskite catalyst for combustion of methane. *Journal of Rare Earths*, Vol. 28, No. 1, pp. 54-58

Zamaro, JM.; Ulla, M.A. & Miro, E.E. (2008). ZSM5 growth on a FeCrAl steel support. Coating characteristics upon the catalytic behavior in the NOx SCR. *Microporous and Mesoporous Materials*, Vol. 115, No. 1-2, pp. 113–122

Zhao, S.; Zhang, J.; Weng, D. & Wu, X. (2003). A method to form well-adhered γ-Al_2O_3 layers on FeCrAl metallic supports. *Surface and Coating Technology*, Vol. 167, No. 1, pp. 97-105

3

Integration of Electrografted Layers for the Metallization of Deep Through Silicon Vias

Frederic Raynal
Alchimer S.A.,
France

1. Introduction

After many years as a hypothetical possibility, 3D integrated circuits (3D IC) stacking has emerged as a potential key enabler for maintaining semiconductor performance trends. Implementing 3D, however, will almost certainly require development of through-silicon vias (TSVs), which in the past few years have been elevated by the semiconductor industry to the status of a crucial mainstream technology.

TSVs sit at the foundation of the 3D-IC revolution and are a key enabler for extending semiconductor integration trends into a new phase. Integrated device manufacturers and fabless design houses need small, high-density, fine-pitch vias for improved signal integrity and Si real-estate savings. They need them now, and cannot wait for very thin wafer processing and handling technologies to become mainstream – TSVs must cope with current mainstream wafer thickness.

Deep TSVs with aspect ratio (AR) greater than 15:1 elegantly fulfill both requirements. But they cannot be manufactured with acceptable yield/cost using traditional dry processes for liner, barrier and seed deposition. Chemical and physical vapor deposition (CVD, PVD) techniques show basic shortcomings and impose high capital investments, holding back the industry-wide adoption of 3D-IC solutions. Beyond that, physical limitations of PVD and CVD techniques prevent reaching a good step coverage of the deposited layers inside the vias (step coverage is defined by the ratio between thickness deposited on top vs. bottom of the vias), which is required to perform void-free gap filling of electroplated copper.

Electrografting (eG) and chemical grafting (cG) (Bureau et al., 1999; Palacin et al., 2004; Pinson & Podvorica, 2005; Voccia et al., 2006; Belanger & Pinson, 2011) are two fundamental molecular engineering technologies, delivering high-quality films for high-AR (HAR) TSVs (Suhr et al., 2008). The term "grafting" indicates the formation of strong chemical bonds at the molecular level between the underlaying layer's extreme surface (e.g., silicon) and the film being grown from the surface out (e.g., the isolation liner).

This nanotechnology solution generates surface-initiated conformal films which are thin, continuous, adherent and uniform. They are wet-process techniques, operated in standard plating tools, which have proven their efficiency to deposit ultra-thin (< 10nm) seed layers on PVD (Haumesser et al., 2003, 2004; Raynal et al., 2006) or atomic layer deposition (ALD) (Shih et al., 2004; Shue, 2006) diffusion barriers for interconnect back end of line (BEOL) applications.

1.1 Electrografting (eG) and chemical grafting (cG) mechanisms

Electrografting is based on surface chemistry formulations and processes. It is applied to conductive and semiconductive surfaces, and enables self-oriented growth of thin coatings of various materials, initiated by in-situ chemical reactions between specific precursor molecules and the surface. Contrary to electrodeposition which requires a potential supply throughout deposition to fuel the redox processes, electrografting is an electro-initiated process which requires a charged electrode only for the grafting step, but not for the thickening. As eG is mainly (but not only) a cathodic process, it can generally be applied to various metallic and semiconducting surfaces without any concern over oxide formation.

Electrografting of vinylic polymers onto conducting surfaces has historically been achieved via a direct electron transfer from the cathode to the electro-active monomers in solution. In this approach, eG occurs when vinylic monomers such as acrylonitrile (AN), methacrylonitrile (MAN), vinylpyridine (VP), and methyl methacrylate (MMA), all members of the family of electron-deficient alkenes, are submitted to reductive electrolysis, with a classical three-electrode setup in an anhydrous organic medium (Deniau et al., 1992a). Strictly anhydrous conditions are required here because the resulting radical-anion, together with the anions that allow the propagation of the growing grafted polymer chains, are highly sensitive to proton sources.

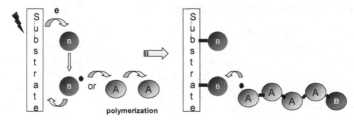

Fig. 1. Schematic representation of radical polymerization electrografting process.

Radical polymerization does not suffer the same drawback and is easily performed in protic conditions (Deniau et al., 2006). Mechanisms of radical polymerization (Tessier et al., 2009) presented in fig. 1 show that polymer electrografting is an electro-induced grafting process followed by a purely chemical propagation step. The first electro-induced step is crucial to form the chemical bond between the polymer and the surface. A specific organic precursor (B) is used both to form a first primer grafted layer and to initiate the polymerization of the vinyl monomer (A) in solution. The termination step of the polymerization leads to the grafting of macromolecular chains (-[A-A-A]$_n$-B) onto the first primer grafted layer.

Chemical grafting is based on the same fundamental mechanisms as electrografting, and is used on any surface (including non-conductive surfaces), electron being replaced by a reducing agent (Mevellec et al., 2007).

1.2 Benefits of electrografting (eG)

Electrodeposition has been known for years for coating conducting surfaces with metals. Electroplating (Kanani, 2005; Schlesinger & Paunovic, 2010) is currently the most widely used method for coating surfaces in many industries (automotive, petrochemical, aerospace) including printed circuit boards, vias and copper interconnections for electronics.

Starting in the 80's with the electrochemical reduction of acrylonitrile on a metallic cathode (Lecayon et al., 1982), electrografting is a relatively new technique, with two major differences compared to electrodeposition: (I) eG being an electro-initiated process, i.e. a process in which Faradaïc electrochemical reactions are coupled to a range of non-Faradaïc chemical reactions, current densities applied for thickening are lower than the current densities required by electrodeposition techniques; and (II) the initial charge transfer to the first monomers leads to the formation of a direct covalent bond between some surface atoms and carbon atoms of the polymeric backbone.

Lower current densities applied are a benefit of electrografting technology when applied to resistive underlaying layers, coatings being more uniform and less sensitive to the ohmic drop involved by underlaying layer resistivity (Gonzalez et al., 2006; Mevellec, 2010).

Direct covalent bonds achieved between coatings and surfaces involve highly adherent electrografted films, via very strong substrate-molecules links, studied by quantum chemistry simulations (Bureau et al., 1996). Experimental evidence of those covalent bonds is not an easy task, and has been demonstrated by XPS analysis technique. As reported in (Deniau et al., 1992b), XPS measurements on very thin eG films exhibit a low-energy shoulder in the C1s signal. This peak cannot be observed for thicker films because it is buried under the strong C1s signals arising from the grafted polymer itself. The corresponding binding energy (283.6 eV) was later attributed to the carbon-nickel bond that links the polymer chain to the Ni electrode (Bureau et al., 1994).

2. Complete metallization of high aspect ratio (HAR) TSVs using electrografting

Aspect ratio is defined by the ratio between the diameter of the via and its depth, and a ratio greater than 10:1 defines the HAR TSVs category. Producing these vertical connections is achieved by: drill a blind hole through the silicon wafer, deposit a uniform liner layer of dielectric material to electrically isolate the via, deposit a barrier layer to prevent copper from diffusing into silicon, and then completely fill the via with electro-chemically deposited (ECD) copper. Chemical mechanical polishing (CMP) and wafer-thinning steps conclude the sequence.

2.1 Comparison between dry and wet TSVs metallization scheme

While the process flow to metallize TSVs is relatively simple, the industry's conventional approach to bringing it into volume production is, in essence, a patchwork of dry process equipment and consumables, such as plasma-enhanced CVD (PE-CVD), ionized PVD (iPVD), and ALD, which were originally designed for dual-damascene applications. One of the main benefits of electrografting is its large reduction in cost of ownership per wafer with respect to conventional dry approaches (Lerner, 2008; Truzzi & Lerner, 2009).

Comparison between dry and wet TSVs metallization scheme is presented in fig. 2. Two schemes are available for electrografting : (1) the deposition of the full stack "wet isolation/ NiB barrier/ Cu seed layer", followed by Cu filling using typical acidic Cu fill chemistry; (2) the deposition of "wet isolation / NiB barrier" followed by Cu filling using a new TSV-grade Cu fill chemistry. Path (2) is preferred because it requires only three steps (instead of four), which involves less cost and higher throughpout.

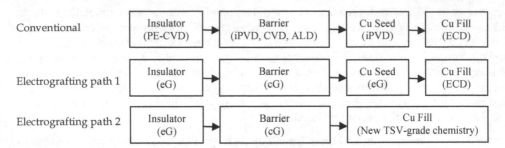

Conventional | Insulator (PE-CVD) → Barrier (iPVD, CVD, ALD) → Cu Seed (iPVD) → Cu Fill (ECD)

Electrografting path 1 | Insulator (eG) → Barrier (cG) → Cu Seed (eG) → Cu Fill (ECD)

Electrografting path 2 | Insulator (eG) → Barrier (cG) → Cu Fill (New TSV-grade chemistry)

Fig. 2. Comparison between dry and wet (electrografting) TSVs metallization scheme.

2.1.1 Description of electrografting path 1 for TSVs metallization

A polymer layer is directly grafted onto the silicon surface, yielding a highly conformal and adherent coating. This first grafted layer acts as an insulating layer (fig. 3a) as well as an adhesion promoter for the subsequent barrier layer deposition, performed by chemical grafting (cG).

Chemical grafting is based on the same fundamental mechanisms as electrografting, and is used on non-conductive surfaces. Specific chemical groups have been chosen to strongly bond the barrier activator with the polymer. This improves adhesion between the barrier and the polymer through a chemical grafting step - it creates a chain of chemical bonds from the substrate to the barrier. The barrier film (fig. 3b) consists of a NiB alloy. Activation of the electrografted insulating layer is carried out at ambient temperature using a metallic catalyst.

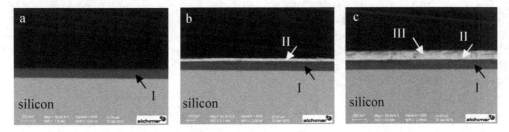

Fig. 3. SEM cross sections of wet insulator (I), NiB barrier (II) and Cu seed layer (III) electrografted on top of silicon surface.

A bath containing specific organics and copper is used to deposit a Cu seed layer (fig. 3c) on the NiB barrier by means of the same electrografting technique. An electrochemical process is applied to provide a conformal and continuous Cu seed layer directly on the NiB barrier.

Both NiB barrier and Cu seed layer can also be deposited on top of dry dielectrics, as for example SiO_2, SiC, SiOC or SiN. SEM cross-sections of barrier and Cu seed deposited on top of SiO_2 are reported in fig. 4.

The electrografted copper seed is also directly applicable to various dry-deposited diffusion barriers, without any adhesion promoter in between (Ledain et al., 2008; Raynal et al., 2009).

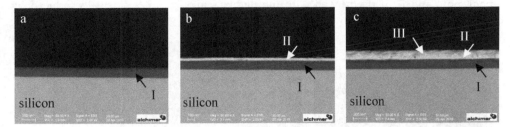

Fig. 4. SEM cross sections of NiB barrier (II) and Cu seed layer (III) electrografted on top of SiO_2 (I) surface.

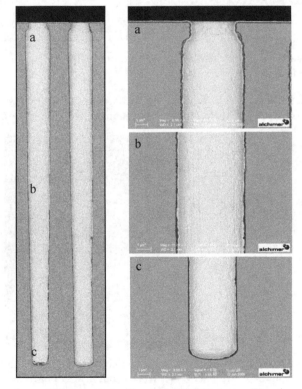

Fig. 5. SEM cross sections of high aspect ratio via (AR 18:1, 4μm x 72μm) coated with electrografted layers; (a) top of via, (b) via middle, (c) via bottom.

Electrografting and chemical grafting formulations and processes fulfill all standard wafer fab requirements and safety guidelines and have been developed and specifically tailored for TSV diameters ranging from 1 to 200 μm, covering a depth/diameter AR range from 2:1 to 20:1 (fig. 5). Higher ARs are possible.

Electrografted and chemical grafted layers, activated from the surface, are not sensitive to its topography, and fit perfectly well with the highly scalloped TSVs sidewalls induced by Bosch etching process (fig. 6).

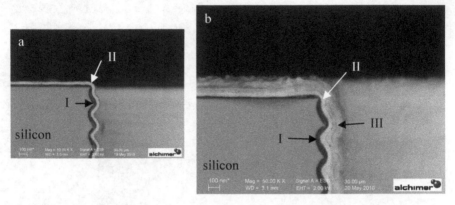

Fig. 6. SEM cross sections of wet insulator (I), NiB barrier (II) and Cu seed layer (III) on top of scalloped silicon surface.

Wet TSVs metallization process is scalable up to 12inch wafers, as illustrated in fig. 7 with three 12inch wafers stopped at different steps of the metallization process.

Fig. 7. Top view of 12inch wafers with various layers electrografted on them; (a) wet insulator; (b) wet insulator / NiB barrier; (c) wet insulator / NiB barrier / Cu seed layer.

2.1.2 Description of electrografting path 2 for TSVs metallization

Wet insulator and NiB barrier deposition remains exatly the same as in path 1.

Filling narrow, deep vias without voids is not an easy task. Most commercialy available chemistries encounter problems due to the sheet resistance (R_s) of the underlying layer, and this is the reason why Cu-seed layers are required. However, because of the extremely poor step coverage of dry-process barriers and seed layers (< 10%), sheet resistance values at the bottom of the via are very high, making it difficult to initiate the filling process.

Via filling completion directly from the barrier requires barrier layers with low resistivity values, and copper plating chemistries with low sensitivity to ohmic drop. NiB barrier has been selected because of its barrier properties (see section 2.2.2.), and because Ni-based barrier films present much lower resistivity values than conventional dry barriers (fig. 8). Indeed, NiB chemical grafting barrier formulations have been optimized to reduce the resistivity value below 25µΩ.cm; this value makes it possible to eliminate Cu seed layers, and sets the stage for direct fill of TSVs from the barrier layer, further simplifying the TSV process sequence.

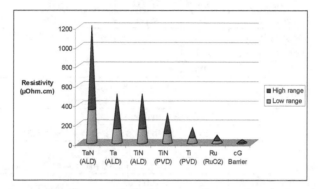

Fig. 8. Resistivity range of conventional dry barriers, compared with chemically grafted (cG) NiB barrier.

Today, a new mildly basic TSV plating chemistry is available to fill the vias (Truzzi, 2010) based on the same nanotechnology concept described in this paper. This TSV-grade chemistry is not sensitive to the sheet resistance of the underlying layer, and can be applied over R_s values up to 50 Ω/sq. It is fully compatible with industry standard wet-process tools and, in contrast with ECD solutions, it does not attack or degrade the underlying layer.

Some examples of copper filling in various TSV dimensions, achieved directly from the NiB barrier, are presented in fig. 9. The new TSV-grade Cu fill chemistry has been used in all cases. Fig. 9a shows an example of copper filling achieved in different TSV dimensions at the same time, which demonstrates the wide process window of this new TSV-grade chemistry regarding vias size.

2.2 Properties of electrografted wet insulator layer, NiB barrier layer, Cu seed layer and new TSV grade Cu fill chemistry

Film thickness can be controlled to any value from 40 to 400 nm with maximum non-uniformity of 5% within wafer (300mm). This provides a step coverage value (bottom/top thickness ratio) up to 90% for liner and metal layers. As a reference, typical dry-process barrier step coverage values are lower than 10% for 10:1 aspect ratio TSVs. Adhesion of all layers was measured using a 16-squares scribe tape test method: all layers successfully passed the test. Film properties of each layer, as well as reliability test results, are discussed in detail in (Truzzi et al., 2009; Raynal et al., 2010). Selected basic film properties are summarized in table 1.

Parameter	Value	Unit	Notes
Wet insulator			
CTE	30	ppm/°C	
Dielectric constant	3		SiO_2 = 4.2
Breakdown Voltage	28	MV/cm	SiO_2 = 10
Capacitance Density	0.13	fF/µm2	
Leakage current	15	nA/cm2	SiO_2 = 10-20
Surface Finish	1.6	Nm	SiO_2 = 2
Substrate resistivity	< 200	Ohm.cm	
Young Modulus	3.4	Gpa	SiO_2 = 107
Stress @ 200nm	10	Mpa	SiO_2 = 100
Moisture Absorption	< 1%		168hrs @ 90%RH
TGA (1hr/temp. point)	450	°C	<4% mass loss
Contact angle	50%		
Chemical purity (at%)	S,F,Ca,Na,K : 0 Cl : 0.2 Organic polymer : 99.8		
NiB barrier			
Resistivity	25	µOhm.cm	TiN = 100-250
Rs uniformity	5	%	
Barrier property	Equivalent to TiN after 400 °C 2 hours		
Cu penetration after 400°C 2 hours	42	% barrier thickness	TaN/Ta = 54
Hardness	14.3	GPa	TiN = 25
Stress	200	MPa	TaN = 1500 TiN = -750
Cu seed layer			
Resistivity	1.8	µOhm.cm	ECD-Cu = 1.8
Rs uniformity	5	%	
Grain size	110	Nm	ECD-Cu = 100
Stress	50	MPa	
ECD gap-fill compatibility	No voids		
New TSV-grade Cu fill			
Resistivity	1.8	µOhm.cm	ECD-Cu = 1.8
Rs uniformity	5	%	
Chemical purity	Cl : 100x less than baseline plating chemistries C : 10x less than baseline plating chemistries S : equivalent to baseline plating chemistries		
Stress	50	MPa	
Copper gap-fill	No voids		

Table 1. Selected film properties of electrografted wet isolation, NiB barrier, Cu seed layer and new TSV-grade Cu fill, compared with industry baseline.

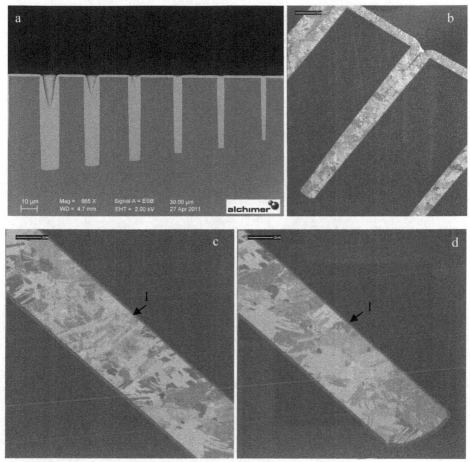

Fig. 9. SEM cross sections of copper filling achieved with the new TSV-grade Cu fill chemistry, directly from NiB barrier; (a) filling of various TSV dimensions (via diameter 3.5µm to 12µm, AR 4.5:1 to 11:1); (b) filling of TSVs 5µm x 50µm (AR 10:1); (c) middle depth of filled TSVs 5µm x 50µm, showing wet insulator layer (I); (d) via bottom of filled TSVs 5µm x 50µm, showing wet insulator layer (I).

2.2.1 Properties of wet insulator

Electrical properties of wet insulator have been characterized using conventional mercury probe analysis tool. Fig. 10 presents a typical C(V) curve recorded with the stack wet insulator / Si p-doped. Dielectric constant calculed from this curve is 3. Breakdown voltage, leakage current and capacitance density have been measured from I(V) curves performed with the same mercury probe analysis method.

Thickness of the wet insulator is measured by ellipsometry. Examples of typical n=f(λ) and k=f(λ) curves recorded with wet insulator are presented in fig. 11. According to n=f(λ) curve, refractory index is measured in the range 1.7-1.8nm.

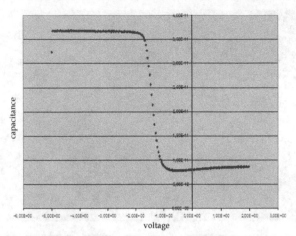

Fig. 10. Typical C(V) curve of the stack wet insulator / Si p-doped.

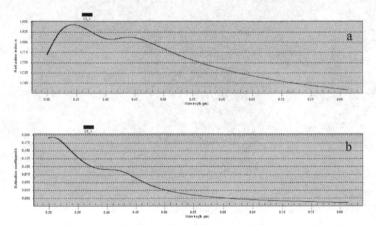

Fig. 11. Typical (a) n=f(λ) and (b) k=f(λ) curves recorded with wet insulator.

CTE (coefficient of thermal expansion) of the wet insulator is measured using an ellipsometer linked to a heater. As presented in fig. 12, a CTE of 30ppm/°C is measured for wet insulator, 30ppm/°C being the mean value obtained on four different samples. On each sample, CTE was measured after one cycle of sample warm and cool.

Wet insulator being a polymer, T_g value can be deducted from CTE curves (Fryer et al., 2001). Indeed, wet insulator thickness evolution vs. temperature being linear from 50°C to 200°C, this means that no microstructural reorganization is observed in this range of temperatures, and T_g value of the wet insulator is higher than 200°C.

Reduced modulus has been measured using nanoindentation analysis technique, with a maximum force applied of 5000 nN and a maximum indenting depth of 35 nm.Typical curve of force vs. penetration depth is presented in fig. 13a, Young modulus characterization being achieved during the loading of the indentor inside the wet insulator. Young modulus measurement of 4.05 GPa is the mean value of 15 measurements (fig. 13b).

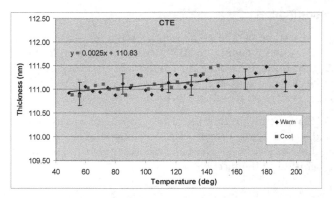

Fig. 12. Coefficient of thermal expansion (CTE) measured with wet insulator.

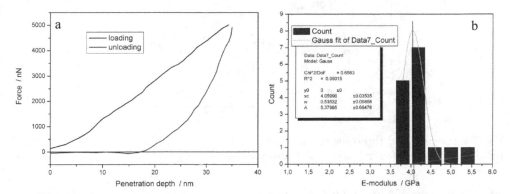

Fig. 13. (a)Typical curve of force vs. penetration depth recorded by nanoindentation analysis of wet insulator; (b) values obtained on 15 measurements.

Thermo-Gravimetry Analysis (TGA) has been performed to check the stability of the wet insulator under thermal stress. Experimental conditions are the following: anneal is achieved under nitrogen atmosphere, for a duration of one hour at each selected temperature. Range of analysis is 300°C – 500°C, with a step of 50°C between each experiment. Fig. 14 shows that less than 1% of mass loss is measured after 350°C applied during one hour to the wet insulator, and less than 4% after 450°C applied during one hour.

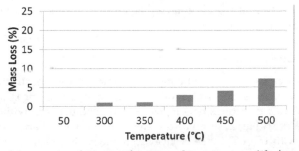

Fig. 14. Mass loss of the wet insulator as a function of temperature (1h / temp. point).

2.2.2 Properties of NiB barrier

Barrier property is measured by TOF-SIMS depth profile, after anneal of the complete stack (wet insulator / NiB barrier / copper) to 400°C during 2 hours under forming gas. Analysis is performed starting from the front side (copper) of the sample, or from the backside (silicon) of it. In the case of a start from the back side, a mechanical polishing and grinding of the sample is first performed to reduce silicon bulk thickness.

Examples of TOF-SIMS depth profiles are shown in fig. 15, x axis being representative of sample depth and y axis being proportional to compound quantity. Barrier property is quantified by the difference between NiB barrier and copper signals. The better is the barrier property, the higher is this difference.

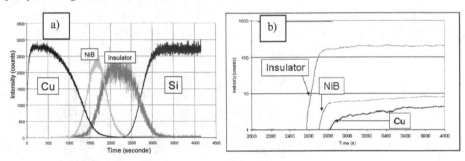

Fig. 15. ToF-SIMS depth Profile of Cu layer (100nm) on top of NiB barrier (40nm) and wet insulator (200 nm) after anneal to 400°C during 2 hours using (a) front side and (b) back side analysis.

As shown in fig. 15b, diffusion of Cu atoms occurred only into the NiB barrier layer. Cu atoms are effectively blocked by this layer, which prevents diffusion into the wet insulator and the silicon substrate. NiB film acts as a very efficient copper diffusion barrier, and a proven surface for subsequent Cu deposition.

Porosity of the barrier has been checked using a method based on the chemical attack of SiO_2 by Hydrofluoridric Acid (HF) solutions. The stack PE-CVD SiO_2 / NiB barrier (200nm / 50nm) has been deposited on top of a Silicon substrate, the whole stack being dipped into HF 1% during 5 minutes. After this chemical treatment, if some porosity is present into the NiB barrier, than HF is going to diffuse in it, and attack the underlaying layer SiO_2. Fig. 16 demonstrates that no attack of the underlaying layer SiO_2 is observed after dipping in HF, and therefore NiB barrier is not porous.

2.2.3 Properties of Cu seed layer

A bath containing specific organics and copper is employed to deposit a Cu seed layer on NiB barrier by means of the same electrografting technique. The electrografted copper seed is also directly applicable to various dry-deposited diffusion barriers, without any adhesion promoter in between. Table 2 summarizes the dry diffusion barriers that have been successfully tested with the eG Cu seed layer technology. No instance of incompatibility between a diffusion barrier and eG Cu seed has been reported so far.

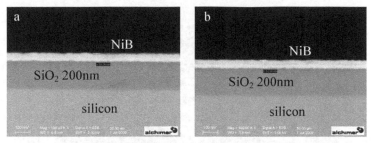

Fig. 16. Demonstration of the non porosity of NiB barrier by dipping the stack SiO_2 / NiB in HF 1%; (a) before HF dip and (b) after HF dip.

	Ta	TaN	Ti	TiN	TiW
ALD				Yes	
CVD				Yes	
PVD	Yes	Yes	Yes	Yes	Yes

Table 2. Compatibility of dry diffusion barriers with electrografting technology.

An example of electrografted copper seed deposited in 5μm x 50μm (AR 10:1) TSVs coated with CVD TiN is presented in fig. 17. Step coverage (step coverage is defined by the ratio between thickness deposited on top vs. bottom of the vias) of eG copper seed deposited inside those TSVs is close to 90%.

Copper grain size of eG copper seed deposited on top of dry barriers had been previously characterized in (Raynal et al., 2009). These measurements were re-visited in the case of eG copper seed deposited on top of eG barrier, with similar results obtained by XRD, EBSD and self anneal rate. One example of EBSD measurement is shown in fig. 18.

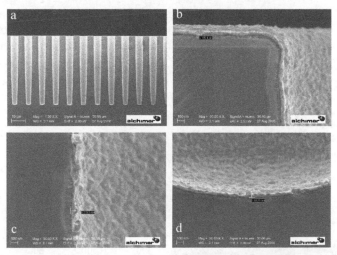

Fig. 17. SEM cross sections of eG copper seed deposited in 5μm x 50μm (AR 10:1) TSVs coated with CVD TiN: (a) overview; (b) via top (109nm of eG copper); (c) sidewalls (129nm of eG copper); (d) via bottom (98nm of eG copper).

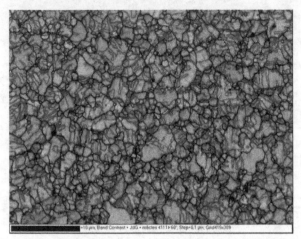

Fig. 18. EBSD picture of eG copper seed layer deposited on top of NiB barrier.

Several studies of PVD copper microstructure have been performed in the past, especially when PVD Cu is deposited on top of PVD Ta. In this case, a strong (111) texture of the copper is reported in the literature, linked to the heteroepitaxial growth of Cu on (002) orientation of Ta. This heteroepitaxial relationship is observed between Cu and Ta, but not with TiN barriers (Wong et al., 1998).

Copper microstructure has been compared between PVD copper deposited on top of TiN barrier and eG copper seed deposited on top of NiB barrier. In both cases, weak (111) orientation of the copper layer is measured.

A summary of eG copper resistivity evolution vs. layer thickness is presented in fig 19. The trend reported in fig. 19 is similar to PVD copper layers (Barnat & Lu, 2001), copper resistivity below 100nm thick being impacted by grain boundary and surface scattering.

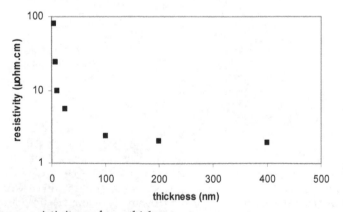

Fig. 19. eG copper resistivity vs. layer thickness.

Thermal treatments need to be done to decrease copper layer resistivity and to reach the value of $1.8\mu\Omega$.cm for layers thicker than 100nm. This phenomenon is related to grain size growth during annealing, which can be observed by XRD or EBSD analysis techniques.

Successful Cu gap fillings have been demonstrated, starting from eG copper seed, in a wide range of TSV dimensions. Those results have been achieved with conventional acidic plating chemistries. SEM cross-sections presented in fig. 20 show some examples of void-free gap fillings, starting from eG copper layers with a step coverage between 60% and 90%, in the case of both 20µm x 75µm and 50µm x 100µm TSVs.

Fig. 20. SEM cross sections of TSVs filled with conventional acidic plating solutions, starting from eG copper seed: (a) 20µm x 75µm TSVs; (b) 50µm x 100µm TSVs; (c) via bottom zoomed 50µm x 100µm TSVs.

2.2.4 Properties of the new TSV-grade Cu fill chemistry

This new TSV-grade chemistry is dedicated to the copper filling of TSV structures. It is less aggressive than conventional acidic chemistries, involving no degradation of the underlying layer, and less sensitive to the ohmic drop, which allows to fill directly on top of the NiB barrier. Its formulation is simpler, which involves a better process control, and a lower contamination of the copper deposited.

Contamination in the copper bulk has been compared by TOF-SIMS depth profiles, between this new TSV-grade chemistry (New fill) and a conventional acidic plating chemistry for TSV applications (Baseline). TOF-SIMS depth profiles are presented in fig. 21, x axis being representative of sample depth and y axis being proportional to compound quantity. Fig. 21 shows an order-of-magnitude reduction in carbon contamination and a two orders-of-magnitude reduction in chlorine contamination for the new TSV-grade copper fill compared to conventional plating chemistries. Those trends remain similar before and after copper anneal to 250°C under forming gas.

Copper grain size has been compared between new TSV-grade and conventional plating chemistries for copper 1µm thick (fig. 22), using EBSD analysis technique. Grain size is higher in the case of new TSV-grade chemistry, and more uniform compared to conventional acidic plating. Stress observed by EBSD in grain boundary mode (fig. 22e & 22f) is lower with the new TSV-grade Cu fill (stress being proportional to red lines density).

Cristalline orientation is also extracted from EBSD analysis. As observed in fig. 23, copper deposited with the new fill chemistry is more (111) oriented than baseline chemistry.

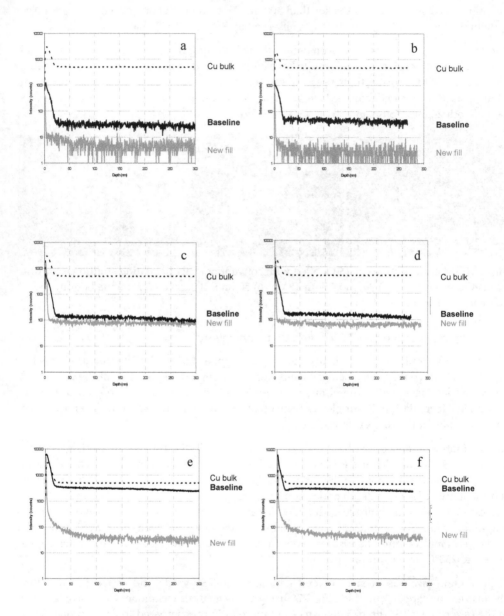

Fig. 21. TOF-SIMS depth profiles of conventional acidic plating (Baseline) vs. new TSV-grade Cu fill (New fill) chemistry: (a) Carbon as deposited; (b) Carbon after anneal 250°C; (c) Sulfur as deposited; (d) Sulfur after anneal 250°C; (e) Chlorine as deposited; (f) Chlorine after anneal 250°C.

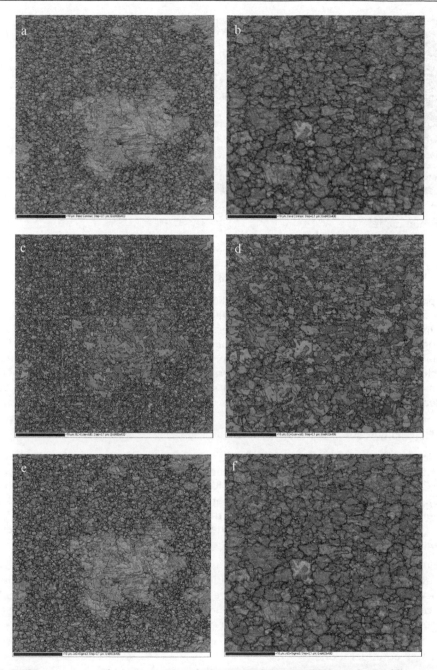

Fig. 22. EBSD top views of (a) conventional plating and (b) new TSV-grade Cu fill in Band Contrast mode; (c) conventional plating and (d) new TSV-grade Cu fill in Euler Angle mode; (e) conventional plating and (f) new TSV-grade Cu fill in Grain Boundary mode.

Fig. 23. EBSD top views of (a) baseline plating and (b) new TSV-grade Cu fill in grain orientation mode.

2.3 Wet process TSV - Reliability

2.3.1 Thermo-mechanical considerations of the stack

As reported in table 1, Young modulus (elasticity module) of the wet insulator is measured below 5GPa, which is much lower than the value obtained with SiO_2 (107 GPa, see table 1). This low value of Young modulus represents an obvious advantage in the use of electrografted layers, related to the mechanical properties of the metallized via. Indeed, thermally-induced stress in bulk silicon is proportional to the square of TSV radius and elasticity module (Lu et al., 2010). Silicon stress is reduced by 30x (97%) when used with the wet insulator. This comes on top of the via size reduction induced by smaller radius, as enabled by electrografting (see 2.4 section).

2.3.2 Reliability tests

Specific test vehicles, with typical TSV design rules, were used to assess the reliability of an integrated stack of electrografting and chemical grafting layers filled with the new TSV-grade Cu fill chemistry. Those tests included temperature cycling (1000 cycles from -55°C to 125°C), moisture sensitivity levels, high-temperature storage, thermal shock, and solder heat resistance (Truzzi, 2010). All samples passed the reliability tests. Fig. 24 shows a SEM cross-section and top view of filled structures after 1000 thermal cycles. Table 3 presents an overview of thermal cycles performed with various stacks.

Additionally, electrografting layers have been integrated into test vehicles and exposed to autoclave (AC) and high-temperature storage (HTS) reliability testing. The autoclave test was conducted for 96 hours at 121°C, 100% relative humidity and 2 bar absolute pressure. High-temperature storage was performed for 20 hours at 205°C. Both tests showed positive results with no significant difference in film performance before and after the tests (Reed et al., 2010).

2.4 Design considerations

For a given TSV depth, this nanotechnology approach allows to manufacture smaller vias, thereby unlocking the possibility to design small, fine-pitch TSVs for demanding via-middle applications such as memory-on-logic for mobile computing. From the signal integrity

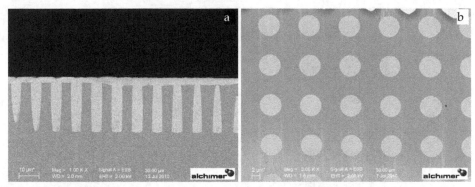

Fig. 24. 5µm x 25µm TSVs coated with wet insulator and NiB barrier, filled with the new TSV-grade Cu fill, and exposed to 1000 thermal cycles from -55°C to 125°C: (a) SEM cross-section and (b) SEM top view after copper polishing.

Via size	Insulator	Barrier	Copper seed	Fill	Conditions	Result
50x100µm	Dry SiO2	Dry TiN	eG Cu seed	Acidic plating	10 cycles from 30°C to 300°C. Ramp up 54°C/min → plateau 300°C 5min → ramp down 33°C/min → plateau 30°C 5min	PASS
50x100µm	Dry SiO2	Dry TiN	eG Cu seed	Acidic plating	1000 cycles from -55°C to 125°C. Ramp up 36°C/min → plateau 125°C 15min → ramp down 36°C/min → plateau -55°C 5min	PASS
5x25µm	Dry SiO2	Dry TaN/Ta	eG Cu seed	Acidic plating	1000 cycles from -55°C to 125°C. Ramp up 36°C/min → plateau 125°C 15min → ramp down 36°C/min → plateau -55°C 5min	PASS
5x25µm	eG wet insulator	cG NiB	-	New TSV-grade	1000 cycles from -55°C to 125°C. Ramp up 36°C/min → plateau 125°C 15min → ramp down 36°C/min → plateau -55°C 5min	PASS
7x25µm	eG wet insulator	cG NiB	-	New TSV-grade	1000 cycles from -55°C to 125°C. Ramp up 36°C/min → plateau 125°C 15min → ramp down 36°C/min → plateau -55°C 5min	PASS
10x30µm	eG wet insulator	cG NiB	-	New TSV-grade	1000 cycles from -55°C to 125°C. Ramp up 36°C/min → plateau 125°C 15min → ramp down 36°C/min → plateau -55°C 5min	PASS

Table 3. Overview of thermal cycles performed with various stacks.

standpoint, as shown in (Truzzi & Lerner, 2009), HAR TSVs with a diameter ranging from 1 to 5µm and a depth ranging from 25 to 100 µm show smaller self-capacitance and less cross-talk than larger TSVs with similar depth (Weerasekera, 2008). Fig. 25 shows how one single large TSV exhibits a worse transmission behavior than nine small properly positioned TSVs, with ground vias correctly shielding signal lines. As for the impact of TSVs on Si real estate, the area needed for vias decreases exponentially with decreasing diameter (fig. 26). In order to illustrate this point, we can consider the assumptions listed in table 4.

Fig. 26 shows an exponential dependency of Si cost on via diameter for constant depth value: savings grow exponentially with increasing aspect ratio. In other words, a technology enabling a 3X improvement in aspect ratio allows a 8X increase in the number of TSVs per given area.

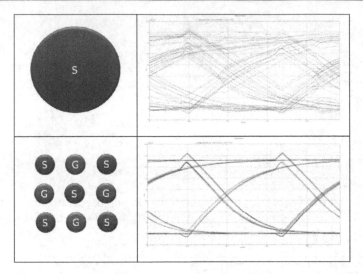

Fig. 25. Eye diagram comparison between one large single TSV and nine small, correctly designed vias.

Die	10 x 10	Mm
die area	100	mm2
# TSVs	10000	
TSV density	100	TSV/mm2
TSV depth	50	μm
Keep out area	2.5X diameter	

Table 4. Si area penalty model.

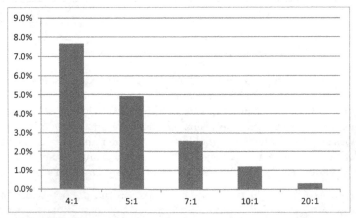

Fig. 26. Si area penalty for TSVs as a function of aspect-ratio, keeping a constant depth value of 50μm.

2.5 Bath analysis and manufacturing

The electrografting nanotechnology described in this paper is ready to be used in a production environment. The electrochemical baths are highly stable, shelf life of all the chemistries being more than 3 months.

After preparation of the chemical baths, monitoring and replenishment of the solutions can be easily carried out using conventional analytical methods, such as pH measurement or UV-visible spectrophotometry. Table 5 summarizes analytical methods available for each bath.

Fig. 27 presents the characterization of the eG copper seed solution during a marathon performed with 900 8inch wafers, using a replenishment of 1ml/h. Process performance (adhesion, copper resistivity, step coverage in TSVs, uniformity within wafer ...) remained stable during the entire marathon.

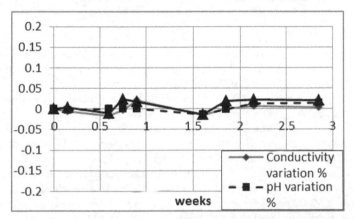

Fig. 27. eG Cu seed bath evolution during a marathon performed with 900 8inch wafers.

	UV-visible	Titration	pH	CVS
Wet insulator	■			
NiB barrier	■	■	■	
eG Cu seed	■	■	■	
New TSV-grade Cu fill	■	■	■	■

Table 5. Analytical methods required for each wet process TSV bath.

3. Conclusion

Electrografting nanotechnology has been optimized for highly conformal growth of TSV films. Complete metallization of high aspect ratio vias is available with this technique, from insulation (by electrografted wet insulator) to copper gap filling using a new TSV-grade Cu fill chemistry.

This technology is fully compatible with standard semiconductor plating tools, enabling a large reduction in cost-of-ownership per wafer compared to dry process approaches (Lerner, 2008), while also providing stable and well-monitored chemical baths.

Film properties meet or exceed current TSV requirements, and chemical formulations are production-ready. TSVs manufactured using electrografting can be very narrow and have aspect ratios up to 20:1, thus broadening the 3D-IC design space and offering a process solution that can be extended for at least several generations into the future.

4. Acknowledgment

The author would like to thank Alchimer S.A. technical team.

eG and cG are trademarks of Alchimer S.A.

5. References

Barnat, E.V. & Lu, T-M. (2001). Real Time Copper Resistivity Measurements during Sputter Deposition, *Proceedings of 4th IEEE International Interconnect Technology Conference*, pp. 24-26, Burlingame, California, USA, June 2001.

Belanger, D. & Pinson, J. (2011). Electrografting: a Powerful Method for Surface Modification, *Chemical Society Reviews*, Vol.40, pp. 3995-4048.

Bureau, C.; Defranceschi, M.; Delhalle, J.; Deniau, G.; Tanguy, J. & Lecayon, G. (1994). Theoretical Monomer/Cluster Model of a Polymer/Metal I: poly(methacrylonitrile) on a Nickel Surface, *Surface Science*, Vol.311, pp. 349-359.

Bureau, C.; Deniau, G.; Valin, F.; Guittet, M-J.; Lecayon, G. & Delhalle, J. (1996). First Attempts at an Elucidation of the Interface Structure Resulting from the Interaction between Methacrylonitrile and a Platinum Anode: an Experimental and Theoretical (Ab Initio) Study, *Surface Science*, Vol.355, pp. 177-202.

Bureau, C. & Delhalle, J. (1999). Synthesis and Structure of Polymer/Metal Interfaces: a Convergence of Views between Theory and Experiment, *Journal of Surface Analysis*, Vol.6, pp. 159-170.

Deniau, G.; Lecayon, G.; Viel, P.; Hennico, G. & Delhalle, J. (1992a). Comparative Study of acrylonitrile, 2-butenenitrile, 3-butenenitrile, and 2-methyl-2-propenenitrile Electropolymerization on a Nickel Cathode, *Langmuir*, Vol.8, pp. 267-276.

Deniau, G.; Viel, P.; Lecayon, G. & Delhalle, J. (1992b). UPS and XPS Study of the Polymer Metal Interface of poly(2-methyl-2-propenenitrile) Electropolymerized on an Oxidized Nickel Surface Cathode, *Surface and Interface Analysis*, Vol.18, pp. 443-447.

Deniau, G.; Azoulay, L.; Bougerolles, L. & Palacin, S. (2006). Surface Electroinitiated Emulsion Polymerization: Grafted Organic Coatings from Aqueous Solutions, *Chemistry of Materials*, Vol.18, pp. 5421-5428.

Fryer, D.S.; Peters, R.D.; Kim, E.J.; Tomaszewski, J.E.; de Pablo, J.J.; Nealey, P.F.; White, C.C. & Wu, W-L. (2001). Dependance of the Glass Transition Temperature of Polymer Films on Interfacial Energy and Thickness, *Macromolecules*, Vol.34, pp. 5627-5634.

Gonzalez, J.; Raynal, F.; Monchoix, H.; Ben Hamida, A.; Daviot, J.; Rabinzohn, P. & Bureau, C. (2006). Electrografting, a Unique Wet Technology for Seed and Direct Plating in Copper Metallization, *ECS Transactions*, Vol.1, pp. 155-162.

Haumesser, P-H.; Giblat, F.; Ameur, S.; Cordeau, M.; Maîtrejean, S.; Mourier, T.; Bureau, C. & Passemard, G. (2003). Electrografting: a New Approach for Copper Seeding or Direct Plating, *Proceedings of 20th Advanced Metallization Conference*, pp. 575-581, Montreal, Canada, October 2003.

Haumesser, P-H.; Cordeau, M.; Maîtrejean, S.; Mourier, T.; Gosset, L.G.; Besling, W.F.A.; Passemard, G. & Torres, J. (2004). Copper Metallization for Advanced Interconnects: the Electrochemical Revolution, *Proceedings of 7th IEEE International Interconnect Technology Conference*, pp. 3-5, Burlingame, California, USA, June 2004.

Kanani, N. (2005). *Electroplating: Basic Principles, Processes and Practice*, Elsevier Science, ISBN 9781856174510, Oxford.

Lecayon, G.; Bouizem, Y.; Le Gressus, C.; Reynaud, C.; Boiziau, C. & Juret, C. (1982). Grafting and Growing Mechanisms of Polymerised Organic Films onto Metallic Surfaces, *Chemical Physics Letters*, Vol.91, pp. 506-510.

Ledain, S.; Bunel, P.; Mangiagalli, P.; Carles, A.; Frederich, N.; Delbos, E.; Omnes, L. & Etcheberry, A. (2008). An Evaluation of Electrografted Copper Seed Layers for Enhanced Metallization of Deep TSV Structures, *Proceedings of 11th IEEE International Interconnect Technology Conference*, pp. 159-161, Burlingame, California, USA, June 2008.

Lerner, S. (2008). Economic Advancement of High-Aspect-Ratio Through-Silicon Vias for 3D Integration, *Future Fab International*, Issue 26.

Lu, K.H.; Ryu, S-K.; Zhao, Q.; Zhang, X.; Im, J.; Huang, R. & Ho, P.S. (2010). Thermal Stress Induced Delamination of Through Silicon Vias in 3D Interconnects, *Proceedings of 60th Electronic Components and Technology Conference*, pp. 40-45, Las Vegas, Nevada, USA, May 2010.

Mevellec, V.; Roussel, S.; Tessier, L.; Chancolon, J.; Mayne-L'Hermite, M.; Deniau, G.; Viel, P. & Palacin, S. (2007). Grafting Polymers on Surfaces: a New Powerful and Versatile Diazonium Salt-Based One-Step Process in Aqueous Media, *Chemistry of Materials*, Vol.19, pp. 6323-6330.

Mevellec, V. (2010). Electrografted Polymer Layers for Insulation of Deep TSV Structures, *Proceedings of 14th Symposium on Polymers for Microelectronics*, Wilmington, Delaware, USA, May 2010.

Palacin, S.; Bureau, C.; Charlier, J.; Deniau, G.; Mouanda, B. & Viel, P. (2004). Molecule to Metal Bonds: Electro-grafting of Polymers onto Conducting Surfaces, *ChemPhysChem*, Vol.5, pp. 1468-1481.

Pinson, J. & Podvorica, F. (2005). Attachment of Organic Layers to Conductive or Semiconductive Surfaces by Reduction of Diazonium Salts, *Chemical Society Reviews*, Vol.34, pp. 429-439.

Raynal, F.; Guidotti, E.; Gonzalez, J.; Roy, S. & Getin, S. (2006). Electrografting of Ultra-thin (sub 10nm) Seed Layers for Advanced Copper Metallization, *Proceedings of 23th Advanced Metallization Conference*, pp. 129-135, San Diego, California, USA, October 2006.

Raynal, F.; Zahraoui, S.; Frederich, N.; Gonzalez, J.; Couturier, B.; Truzzi, C. & Lerner, S. (2009). Electrografted Seed Layers for Metallization of Deep TSV Structures, *Proceedings of 59th Electronic Components and Technology Conference*, pp. 1147-1152, San Diego, California, USA, May 2009.

Raynal, F.; Mevellec, V.; Frederich, N.; Suhr, D.; Bispo, I.; Couturier, B. & Truzzi, C. (2010). Integration of Electrografted Layers for the Metallization of Deep TSVs, *Journal of Microelectronics and Electronic Packaging*, Vol.7, pp. 119-124.

Reed, J.D.; Goodwin, S.; Gregory, C. & Temple, D. (2010). Reliability Testing of High Aspect Ratio Through Silicon Vias Fabricated with Atomic Layer Deposition Barrier, Seed Layer and Direct Plating Deposition and Material Properties Characterization of Electrografted Insulator, Barrier and Seed Layers for 3-D Integration, *Proceedings of 2th IEEE 3D System Integration Conference*, Munich, Germany, November 2010.

Schlesinger, M. & Paunovic, M. (2010). *Modern Electroplating 5th Edition*, John Wiley & Sons, ISBN 9780470167786, Hoboken.

Shih, C.H.; Su, H.W.; Lin, C.J.; Ko, T.; Chen, C.H.; Huang, J.J.; Chou, S.W.; Peng, C.H.; Hsieh, C.H.; Tsai, M.H.; Shue, W.S.; Yu, C.H. & Liang, M.S. (2004). Direct Plating of Cu on ALD TaN for 45nm-node Cu BEOL Metallization, *Proceedings of 50th IEEE International Electron Device Meeting*, pp. 337-340, San Francisco, California, USA, December 2004.

Shue, W.S. (2006). Evolution of Cu Electro-Deposition Technologies for 45nm and Beyond, *Proceedings of 9th IEEE International Interconnect Technology Conference*, pp. 175-177, Burlingame, California, USA, June 2006.

Suhr, D.; Gonzalez, J.; Bispo, I.; Raynal, F.; Truzzi, C.; Lerner, S. & Mevellec, V. (2008). Through Silicon Via Metallization: a Novel Approach for Insulation/Barrier/Copper Seed Layer Deposition Based on Wet Electrografting and Chemical Grafting Technologies, *Proceedings of Materials Research Society Fall Meeting*, pp. 247-255, Boston, Massachusetts, USA, December 2008.

Tessier, L.; Deniau, G.; Charleux, B. & Palacin, S. (2009). Surface Electroinitiated Emulsion Polymerization (SEEP): a Mechanistic Approach, *Chemistry of Materials*, Vol.21, pp. 4261-4274.

Truzzi, C.; Raynal, F. & Mevellec, V. (2009). Wet-Process Deposition of TSV Liner and Metal Films, *Proceedings of 1st IEEE 3D System Integration Conference*, San Francisco, California, September 2009.

Truzzi, C. & Lerner, S. (2009). Electrografting – Unlocking High Aspect Ratio TSVs, *Future Fab International*, Issue 31.

Truzzi, C. (2010). New Generation of Cost-effective Seedless Technologies for Through Silicon Vias, *Proceedings of 20th Asian Session Advanced Metallization Conference*, Tokyo, Japan, October 2010.

Voccia, S.; Gabriel, S.; Serwas, H.; Jerome, R. & Jerome, C. (2006). Electrografting of Thin Polymer Films: Three Strategies for the Tailoring of Functional Adherent Coatings, *Progress in Organic Coatings*, Vol.55, pp. 175-181.

Weerasekera, R. (2008). System Interconnection Design Trade-offs in Three-Dimensional (3D) Integrated Circuits, PhD Thesis, *The Royal Institute of Technology (KTH)*, Kista (S), December 2008.

Wong, S.; Lee, H.; Ryu, C.; Loke, A. & Kwon, K. (1998). Effects of Barrier/Seed Layer on Copper Microstructure, *Proceedings of 15th Advanced Metallization Conference*, pp. 53-54, Tokyo, Japan, September 1998.

4

Fabrication of InGaN-Based Vertical Light Emitting Diodes Using Electroplating

Jae-Hoon Lee[1] and Jung-Hee Lee[2]

[1]*GaN Power Research Group, R&D Institute, Samsung LED Company Ltd., Suwon,*
[2]*Electronic Engineering & Computer Science, Kyungpook National University, Daegu,*
Korea

1. Introduction

Group III–nitride semiconductors and their ternary solid solutions are very promising as the candidates for both short wavelength optoelectronics and power electronic devices (Nakamura et al. 1997; Nakamura et al. 1995; Lee, et al. 2010; Youn, et al. 2004). The AlGaN/GaN heterostructure field effect transistors (HFETs) have a great potential for future high-frequency and high-power applications because of the intrinsic advantages of materials such as wide bandgap, high breakdown voltage, and high electron peak velocity (Smorchkova et al. 2000). Major developments in wide-gap III-nitride semiconductors have led to commercial production of high-brightness light emitting diodes (LEDs). The InGaN-based LEDs have already been extensively used in full-color displays, traffic displays, and other various applications such as projectors, automobile headlights, and general lightings. In particular, white LEDs based on InGaN/GaN quantum well heterostructure are regarded as the most promising solid-state lighting devices to replace conventional incandescent or fluorescent light. It is required, however, to reduce the high dislocation density due to large lattice mismatch between GaN epitaxial layer and sapphire substrate and to increase the light extraction efficiency for fabricating high-performance LEDs. Epitaxial lateral overgrowth (ELOG) or pendeo epitaxy (PE) has been used to partially overcome the problems (Sakai et al. 1997; Zheleva et al. 1997). Although the ELOG and PE process can dramatically decrease dislocation density, the related growth process is complicated and time consuming. Recently, it has been reported that one can not only reduce the threading dislocation density in GaN films, but also enhance the light extraction efficiency by using patterned sapphire substrate (PSS) (Yamada, et al. 2002; Tadatomo, et al. 2001; Wuu, et al. 2006). The PSS technique has attracted much attention for its high production yield due to the single growth process without any interruption. Future demands for these and more advanced applications will require higher light output with lower power consumption. The refractive index of nitride films (n = 2.4) is higher than that of air (n = 1) and sapphire substrate (n = 1.78). The critical angle of the escape cone is about 23°, which indicates that about only 4% of the total light can be extracted from the surface (Huh, et al. 2003). Most of the generated lights in the active layer is absorbed by the electrode at each reflection and gradually disappear due to total internal reflection. In order to improve the light extraction efficiency, it is very important to find the escape cone which the photons generated from LEDs experience multiple opportunities. Several previous works have reported that one can increase the light extraction efficiency of nitride-based LEDs by roughening the sample surface. However, the surface texturing is difficult in conventional LED because the p-GaN top layer being too thin for texturing and the sensitivity of p-GaN to plasma damage and electrical deterioration. It is also well known that poor heat dissipation has been the main problem in the development of high-power In-GaN/sapphire LEDs. For a conventional nitride-based LED, the generated heat is dissipated through the path from sapphire to silver paste and then to heat sink. The poor thermal conductivities of sapphire ($\sim$35 W/mK) and silver paste (2-25W/mK) limit heat flow and dissipation, resulting in increased junction temperature of the device due to Joule heating at the p-n junction. Recently, high efficiency GaN based LEDs was fabricated using the laser lift-off (LLO) technique (Fujii, et al. 2004; Wang, et al. 2005; Wang, et al. 2006). The LLO process can enhance the light output power, the operating current and the heat sink of fabricated vertical type LED (VT-LED). The other advantage of VT-LED is room for roughening top surface compared to conventional lateral LED. It has been reported to produce a roughened VT-LED surface with conlike feature and microns arrays using photoelectrochamical etching and plasma etch process, respectively. However, since these approaches involve difficult and complicated fabrication process, improvement is still required for higher device performance. In this work, we have proposed a new method utilizing surface texture of vertical type LEDs by using cone-shape patterned sapphire substrate (CSPSS) for the purpose of increasing extraction efficiency (Lee, et al. 2006; Lee, et al. 2008; Lee, et al. 2009) Compared to conventional roughening process, the proposed method is simple and highly reproducible due to forming patterns neatly designed on sapphire before GaN growth. In this chapter, firstly the history and important properties of the group III-nitrides is briefly reviewed, then the fabrication of the InGaN-based VT-LED was presented, finally a conclusion is drawn.

2. Evolution of III-nitride materials

GaN was first created in 1932 by Johnson at the University of Chicago by reacting ammonia gas with solid gallium at temperature of 900-1000°C (Johnson et al. 1932). The GaN synthesis technique was also used to study the crystal structure and lattice constant by Juza in 1938 and to study the photoluminescence spectra by Grimmeiss in 1959 (Juza, et al. 1938; Grimmeiss, et al. 1959). Maruska and Tietjen deposited the first large area GaN layers on sapphire in 1969 using vapor-phase transport in chlorine (Maruskam, et al. 1969). This earliest research on GaN was directed at understanding the properties of the material and learning how to design the material for specific uses. All the early GaN was very conductive as grown, and it was realized that these crystals were n-type with high carrier concentration owing to the highly defective nature of the crystal (Pankove, 1997). Because the work on GaN had early been focused on its potential for optical sources, p-type GaN was required in order to make p-n junction emitters, so much effort was focused in this area. Pankove et al. were successful in fabricating the first blue GaN LED (Pankove, et al. 1972). It was as metal/insulating GaN:Zn/n-GaN (MIS) diode. The n-GaN layers contained different concentration of Zn. The wavelength of the emitted light depended on the Zn concentration. Until the late 1970s the quality of GaN was not very good. The results of optical and electrical measurements were not reproducible. Yoshida et al. deposited an AlN buffer layer to overcome the nucleation problems of GaN grown directly on a sapphire substrate in 1983 (Yoshida, et al. 1983). This buffer layer technique was very effective in improving the overall material quality on sapphire. The two-step method was investigated and perfected by Akasaki, Amano and co-worker in 1988/1989 (Amano, et al. 1986). In the first step a thin buffer layer of AlN is grown on the sapphire substrate at a low temperature of ~ 500°C. The GaN layer is grown on the buffer layer. This importance accomplishment allowed GaN to be grown with much higher crystal quality than was previously available. The world wide attention on the research of group III-Nitrides were stimulated by the success in the material quality improvement with low temperature GaN nucleation layer by Nakamura in the early 1990's and later by the success in blue LEDs and LDs by Nakamura, leading toward the fundamental understanding of group III-Nitride material system (Nakamura, 1991; Nakamura, et al. 1996). The GaN-based FETs were also fabricated about the same time, i.e., in 1993 (Khan, et al. 1993). Fabrication of a heterostructure bipolar transistor using p-type 6H SiC as the based and n-type GaN as the emitter was reported in 1994 (Pankove, et al. 1994). Metalorganic chemical vapor deposition (MOCVD) has been the most popular growth technique for nitride growth and has been the choice for several electronic and optoelectronic devices such as HEMTs, HBTs, LEDs, and LDs.

2.1 Important properties of III-nitride materials

The III-nitride group of materials includes GaN, AlN, InN, and their alloys. The bandgap of the nitride group of materials is direct and have different band gaps ranging from 6.2 eV for AlN down to 0.7 eV for InN. This gives a tremendous optical range for III-nitride materials, from yellow to ultraviolet. The large bandgap and strong termodynamical stability of the III-nitride material system point to promising high power and high-temperature electronic device applications. Nitride materials are also extremely resilient to harsh chemical environments, giving the possibility for use in environments where the more traditional III-V materials are unable to perform. The most thermodynmically stable form of the nitride materials is the wurtzite or hexagonal form. This result in the familiar hexagonal close pack (HCP) model of alternating planes of Ga and N atoms stacked in an ABABAB sequence. The crystal planes in the wurtzite system are given using the Miller-Bravis index system of [ijkl] with k = - (i + j). Further detail on HCP and hexagonal crystal notation can be found in the classic textbook "Introduction to Solid State Physics" by Charles Kittel (Kittel, 1995). Nitrides can also be forced to grow in a zincblend form by using cubic substrates such as (001) Si and GaAs. The stacking sequence in this case is ABCABC. However, although the electrical properties of cubic III-nitrides are in theory expected to be better than for hexagonal GaN, in practice cubic III-nitrides are generally not as high quality as the wurtzite form and often contain a mixture of both cubic and wurtzite phases (mixed phase). The structural properties of the cubic nitrides can be found in Ref (Gil, 1998).

The most common substrate for GaN epitaxial growth is sapphire. It was the substrate used during the first attempts at GaN epitaxy during the late 1960's and has remained as the leading choice to this day. The attraction to sapphire for GaN epitaxy stems from its high melting point, chemical stability, availability of large wafer sizes, very high quality crystals, and low wafer cost. However, sapphire has a large lattice and thermal mismatch with GaN, leading to high defect density (~10^9/cm^2) in the GaN film and limiting the maximum layer thickness to less than 10 μm. In addition, the low thermal conductivity of sapphire limits heat dissipation through the substrate, imposing severe limits for power devices on this substrate. Despite these difficulties, sapphire remains as leading candidate for basic of III-nitride epitaxy due to wide availability and low cost. In the case of growth on c-plane sapphire, a 30° rotation around the c axis occurs for GaN Epitaxy, resulting in an effective lattice constant that is $1/\sqrt{3}$ less than a sapphire. The resulting lattice mismatch is ~ 13%. In addition, on c-plane sapphire, the thermal expansion mismatch is -25.5 %.

Table 1 lists some of the common electronic properties of the III-nitride group of materials (for comparison, properties of Si, GaAs, and SiC are also included) (Levinshtein, et al. 2001). As can be seen, both SiC and the nitride group have larger bandgap energy than either Si or GaAs. The larger bandgap and high breakdown field of these materials allow SiC and Nitride based electronic devices to operate at higher temperatures and higher powers than Si or GaAs based devices. In addition, while SiC has a badgap near that of GaN, the III-nitrides group has three distinct advantages over SiC devices. First, the nitride groups of materials have direct transition bandgap, allowing for use in optical devices such as lasers and LEDs. Secondly, the nitride group has the availability of ternary (AlGaN, InGaN) and quaternary (AllnGaN) compounds leading to a more versatile bandgap range (0.7~6.2) than SiC alone and allowing for bandgap engineering in both optical and electronic application. Finally, the nitride ternary and quaternary compounds can be used for heterojunction

Material	Bandgap (eV)	Lattice constant (Å)	Lattice mismatch with GaN(%)	Thermal Conductivity at 300K (W/cmK)	Thermal expansion coeff. (10^{-6}/K)
AlN	6.2	a=3.112 c=4.982	2.48	2.0	4.2 5.3
GaN	3.44	a=3.1891 c=5.1855	0	1.3	5.59 3.17
InN	0.7	a=3.548 c=5.7034	-10.12	0.8	5.7 3.7
6H SiC	3.03	a=3.081 c=15.117	3.51	4.9	4.2 4.68
Al$_2$O$_3$	9.0	a=5.431 c=12.982	13.9	0.3	7.5 8.5
Si	1.12	a=5.431	-16.96	1.3	3.9
GaAs	1.42	a=5.653	-20.22	0.5	6.7

Material	Electron mobility (cm^2/Vs)	Hole mobility (cm^2/Vs)	Saturated drift velocity (10^7cm/s)	Breakdown field (MV/cm)	Melting Point(°C)
AlN	135	14	1.4	4-12	3124
GaN	300-1000	<200	2.5 (calc) 2.0 (exp)	5.0	2518
InN	1000-1900	-	2.5	5.0	1873
6H SiC	400	50	2.0	3-5	>2000
Al$_2$O$_3$	-	-	-	-	2015
Si	1500	450	1.0	0.3	1412
GaAs	8500	400	1.0	0.4	1240

Table 1. Crystal and electrical properties of III-nitrides, SiC, Si and GaAs (Kittel, 1995; Gil, 1998; Levinshtein, et al. 2001).

(HFET, HBT) and quantum well based devices. SiC does not have any alloys systems, relegating use of SiC to native oxide based devices (MOSFETs). As seen in Table 1, the undoped electron and hole mobility of the III-nitrides and SiC are relatively low compared to Si and GaAs. The electron mobility for unintentionally doped (UID) single GaN layers on either sapphire or SiC average about 300~400 cm^2/Vs. Undoped mobility in these materials are normally limited by defect related scattering of defect hoping (Ng, et al. 1998). The highest reported mobility for MOCVD grown GaN was ~ 900 cm^2/Vs (Nakamura, et al. 1997). The maximum calculated phonon-limited mobility in GaN is ~1350 cm^2/Vs for electrons and ~200 cm^2/Vs for holes (Look, et al. 2001). Light n type doping of GaN (~1 × 10^{17}/cm^3) results in improved mobility due to screening of dislocations and decreased defect-related conductivity. Further doping of the material leads to mobility degradation due to ionized impurity scattering. It should be noted that much higher mobility is possible (1500~2000 cm^2/Vs) in GaN heterostructures by forming a high mobility two dimensional electron gas (2DEG) channel. Another important property for high frequency operation is the saturation electron drift velocity. This parameter gives the average electron velocity at high electric fields. The high saturated drift velocity of III-nitrides leads to applications at high frequencies than can be achieved from Si. As seen in Table 1, the calculated value of GaN electron velocity is 2.5 × 10^7 cm/s, although experimental evidence is closer to 2.0 × 10^7 cm/s (Wraback, et al. 2000) in bulk GaN and near 1.0 × 10^7 cm/s for sub-micron HFET devices (Ridley, et al. 2004). The discrepancy between theory and experimental velocity saturation was accounted for in bulk GaN as due to high defect density not accounted for in theory and in case of HFET devices as due to hot carrier effects. In addition, the wide bandgap and high breakdown field give the possibility or nitride application at both high-frequency and much higher power levels than Si or traditional III-V materials.

3. Fabrication of InGaN-based vertical type light emitting diodes

The epitaxial layers were grown both on CSPSS and conventional sapphire substrate (CSS) by using a metal organic chemical vapour deposition (MOCVD). Figure 1 shows a schematic layer structure and procedure of fabricated VT-LEDs employing selective Ni electroplating and LLO techniques. The LED structure consists of an undoped GaN layer, a Si-doped n-GaN layer, five layers of InGaN/GaN multiple quantum well (MQW), and an Mg-doped p-GaN layer. The grown sample was patterned with 230 × 640 μm^2 size by a standard photolithographic process. The full mesas were defined by ICP-RIE from p-GaN to sapphire bottom to prevent GaN film from breaking due to different stress between sapphire and GaN film for LLO process. A metal scheme of Ni/Ag/Pt was deposited onto the p-GaN top surface to serve as p-ohmic contact and reflector metal. A TiW and Ti/Au layers were subsequently deposited as a barrier layer and an adhesive layer for the nickel electroplating. To sustain the remaining thin LED structure after the removal of sapphire substrate, the thickness of a nickel layer is around 80 μm. To remove the sapphire substrate, a 248 nm KrF excimer laser with pulse width of 25 ns was used for LLO process. Finally, n-contact pad region was defined by ICP-RIE to expose the n-GaN layer and Cr/Au layer was then deposited on n-GaN without additional semitransparent contact layer. For the characterization of the surface morphology of the grown layers, we used a scanning electron microscope (SEM). The crystal quality of the grown films was investigated by both X-ray diffraction (XRD) and transmission electron-microscopy (TEM). The output power of the fabricated VT-LEDs was measured under DC bias at room temperature.

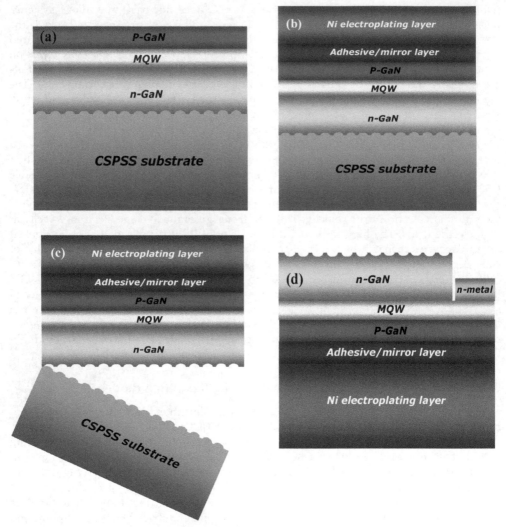

Fig. 1. Schematic layer structure (a) and procedure of fabricated VT-LEDs employing selective Ni electroplating and LLO techniques. (b) p-GaN ohmic contact/mirror layer and Ni electroplating processing, (c) Samples at the LLO processing stage. (b) Samples at the n-GaN ohmic contact processing.

3.1 GaN film on patterned sapphire substrate

Figure 2 shows a scanning electron microscope (SEM) image of the fabricated CSPSS process. The preparation of the CSPSS is as follows [Lee, et al. 2008]. After a photoresist (PR) with 3.5-μm thickness had been coated on a c-plane (0001) sapphire substrate, the PR was patterned first to be a rectangular shape [Fig. 2(a)] with different interval and reflowed during a hard-baking process at 140 °C to make a cone shape as shown in Fig. 2(b). The sapphire substrate was then etched by using inductively coupled plasma reactive ion etching employing reactive Cl_2 gas. The diameter and interval of each cone-shaped pattern were 3 and 1 μm, respectively. The height of the cone shape was about 1.5 μm. The epitaxial layers were grown both on CSPSS and conventional sapphire substrate (CSS) by MOCVD. In the initial stage of the growth on CSPSS, the GaN layer starts to grow only on the etched flat basal sapphire surface, quite differently from the growth mode on the conventional PSS, because there is no preferential growth plane on the cone-shape-patterned region. Similarly to the ELOG, this selective growth on CSPSS prevents the dislocation generated during the initial stage of the growth from propagating further into the patterned area when the growth proceeds laterally toward the cone region, decreases the dislocation density, and hence improves the crystal quality of the grown film.

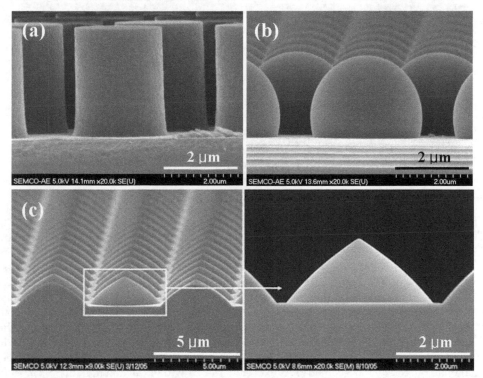

Fig. 2. SEM images of the fabricated CSPSS process. (a) PR patterning, (b) reflow process, (c) fabricated CSPSS.

The time evolution of interference micrographs for a GaN surface grown on a CSPSS is shown in Fig. 3, i.e., after the growth of the initial buffer layer at low temperature [Fig. 3(a)], after temperature ramping (annealing the buffer layer) [Fig. 3(b)], and after 10- and 30-min growth of the epitaxial layer [Fig. 3(c) and (d)]. In general, the III-nitride layer preferentially grows on the (0001) crystallographic c-plane of a sapphire substrate. Although there is no growth plane in the cone-shape-patterned region, the low-temperature GaN amorphous buffer layer is well formed on the plane and patterned sapphire region as shown in Fig. 3(a).

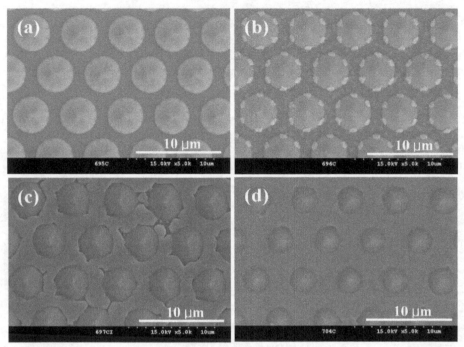

Fig. 3. Time evolution of interference micrographs for a GaN surface grown on a CSPSS. (a) buffer, (b) annealing, (c) 10 min., (d) 30 min.

When the temperature increases from low temperature (550 °C) to high temperature (1020 °C) in the ramping stage of the growth, the random recrystallized GaN islands are formed on the etched flat basal sapphire surface, which is similar to a typical growth mode on the CSS. In the cone-shape-patterned region, however, the larger regular GaN islands are formed on the six corners of a hexagon-shaped cone as shown in Fig. 3(b), which explains that Ga species tend to migrate from the top region of the cone surface to the bottom region of the cone as increasing temperature (Simeonov, et al. 2006; Sugiura, et al, 1997; Sumiya, et al. 2003). In the initial stage of the high-temperature growth on the CSPSS, therefore, the coalescence starts on the etched flat basal sapphire surface, indicating that the growth mode on the CSPSS is considerably different from that on the conventional PSS such as stripe, hexagonal, or rectangular geometry. As expected, the growth of GaN on the CSPSS was only initiated from the etched basal surface with the (0001) crystallographic plane because there was no growth plane in the cone-shape patterned region as shown in Fig. 3(c). As the growth proceeds, the growth also laterally propagates toward the peak of the cone as shown in Fig. 3(d). This lateral growth greatly decreases the dislocation density in the grown film, similarly to the ELOG mode. This also enables reducing the time required in obtaining a smooth growth surface over the patterned region, compared to that for those grown on a conventional PSS, where the growth starts both on the etched and nonetched regions at the same time as shown in Fig. 4. Figure 5 (a) shows the cross-sectional TEM images under $g =$ 0002 two-beam condition of the interface region between the CSPSS and a GaN layer grown on it, demonstrating that the ELOG-like mode on the CSPSS effectively suppresses the propagation of dislocation into the cone region, even though many dislocations were observed in the film grown on the basal plane of the sapphire. This reduction of dislocation was also confirmed by performing a cathodoluminescence (CL) measurement at room temperature as shown in Fig. 5(b, c). In bright regions, a radiative process dominates over a nonradiative process because of the lower density of structural defects. The dark spot density in the film grown on a CSS is roughly estimated to about 7×10^8 cm^{-2}. Apparently, the dark spot density in the film grown on a CSPSS was decreased to about 2×10^8 cm^{-2} in number.

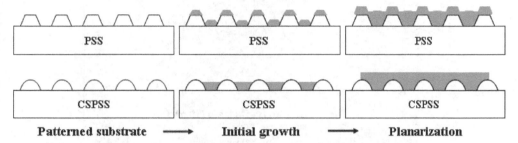

Fig. 4. Schematic view of the planarization of GaN grown on the conventional PSS and on the proposed CSPSS.

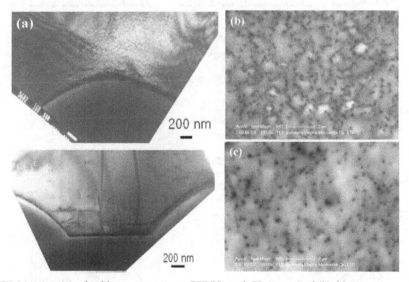

Fig. 5. TEM image (a) of a film grown on a CSPSS and CL image of (b) film grown on a CSS and (c) film grown on a CSPSS.

3.2 Characteristics of GaN films on CSPSS

Room temperature photoluminescence (PL) spectra for both InGaN/GaN samples grown on CSS and CSPSS are shown in Fig. 6. The band-edge emission intensity of the sample grown CSPSS was about four times higher in magnitude than that of the sample grown on CSS. The full width half maximum (FWHM) values of samples gown on CSS and CSPSS are 19 and 17

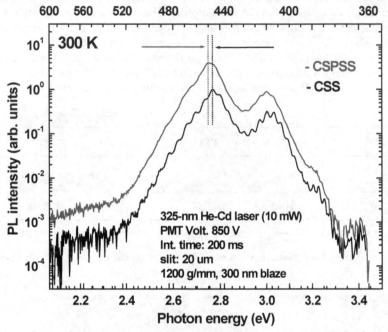

Fig. 6. PL measurements of InGaN/GaN films grown on a CSS and a CSPSS.

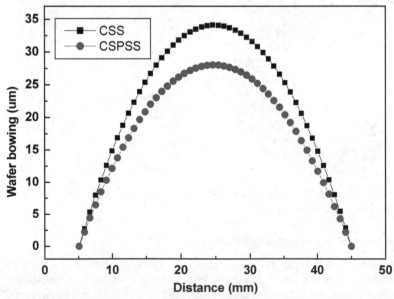

Fig. 7. Wafer bowing of films grown on CSS and CSPSS.

nm, respectively. In conjunction with a considerable enhancement in emission intensity and FWHM in PL, it is clear that the crystal quality of the film grown on CSPSS is improved. The typical corresponding spectral PL peak wavelength from the InGaN/GaN MQWs was shifted from 448 nm for film grown on CSS to 451 nm for film grown on CSPSS. In our previous work (Lee, et al. 2008), we reported that the surface geometry of CSPSS may be adequate for the easy relaxation of compressive strain during the growth on sapphire. The lattice constant c of films grown on CSPSS is very close to lattice constant of the bulk GaN of 5.185 Å, while the lattice constant of the GaN film grown on CSS exhibits a relatively large lattice constant. The slight red shift in PL for the film grown on CSPSS is due to the reduced band-gap energy which is caused by the decreased lattice constant c. The reduction of compressive stress in the films grown on CSPSS was also confirmed by thin film stress measurement using FLX-2320S as shown in Fig. 7. The bowing and stress of the 5.12 μm-thickness film grown on CSS were 34 μm and -437 Mpa with sapphire Young's modulus of 4.08 MPa. On the other hand, the values were reduced to 28 μm and -364 Mpa for the film with thickness of 5.48 μm grown on CSPSS. Consequently, the ELOG-like mode for the GaN layer grown on CSPSS results in less lattice mismatch and incoherency between the GaN layer and the sapphire substrate.

3.3 Role of electroplating

High power GaN-based LEDs are essential for next-generation lighting applications. Essentially, maintain the effective power conversion efficiency (λ=Po/Pe, i.e., the ration of the light output power Po to the electrical power Pe) at a high injection current density is the key to boost their development. However, it is inevitable that the excessive Joule heating from the inherent parasitic resistance and current crowding due to the insulating sapphire substrate in the operation region would eventually decrease radiative recombination efficiency and in turn decrease light output power conversion efficiency. To resolve this situation, the use of a metallic substrate with laser lift-off technology to realize VT-LEDs has been proposed and has attracted much attention (Fujii, et al. 2004; Wang, et al. 2005; Wang, et al. 2006).

3.4 Patterned Ni electroplating

The electroplating process was conducted under a current of around 1.65A with the plating solution kept at about 55 °C, which contains nickel sulphate, nickel chloride, boric acid, water, and some additives to improve surface roughness. The growth rate of nickel layer is about 1000 nm/min. Our experimental results reveal that a nickel layer with a thickness of about 60-80 μm is mechanically strong enough to sustain stress after the removal of sapphire substrate. Figure 8 shows the bowing of film after normal Ni electroplating. The curvature of the bowing increases as thickness of Ni electroplating is increases. This is because the residual compressive stress in the GaN layer, caused by different thermal expansion coefficients between the GaN film and the Ni electroplating substrate. Therefore, the appropriate condition of electroplating is required to prevent the bowing of film. To obtain a flat surface, a patterned Ni electroplating was proposed. Figure 9 shows the schematic view of patterned Ni electroplating. The stress between the GaN and the Ni electrode films was released by using the patterned electroplating through thicker photo resistor (PR), which is resulted in flat film as shown in Fig. 9 (b).

Fig. 8. The bowing of films after normal Ni electroplating.

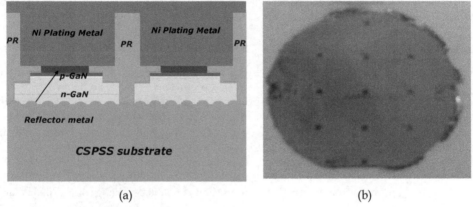

(a)   (b)

Fig. 9. Proposed patterned Ni electroplating (a) schematics of patterned Ni electroplating, (b) The bowing of films after patterned Ni electroplating.

3.5 Concaved surface transfer after Laser lift-off

Figure 10 shows the formation of concaved surface transferred from CSPSS after LLO. To remove the sapphire substrate, A 248 nm KrF excimer laser with a single pulse of 25 ns. The incident laser fluence was set to a value of about $0.75J/cm^2$ which was based on our earlier established threshold laser fluence by taking into account the attenuation of sapphire and the reflection in the sapphire/GaN interface. The laser beam with a size of 1 mm × 1mm was incident from the polished backside of the sapphire substrate into the sapphire/GaN interface to decompose GaN into Ga and N_2. Figure 10 (b), (c) show the SEM images of top surface of CSPSS substrate and removed GaN films after the LLO process. A finite amount of Ga residues appears on the top surface of the GaN and CSPSS substrate. The remaining Ga droplets on the transferred GaN surface were removed by a wet chemical etching using diluted $HCl:H_2O$ (1:2) solution for 1 min. Figure 10 (d) shows the SEM photographs of top surface of removed GaN after the LLO process.

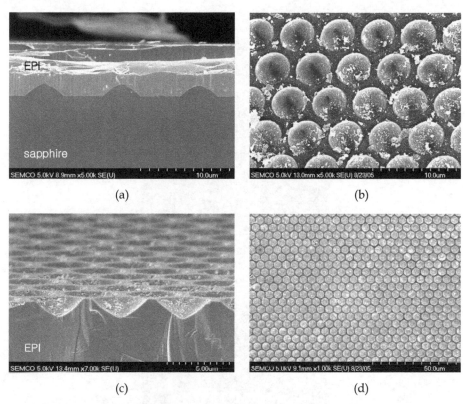

(a) (b)

(c) (d)

Fig. 10. SEM photographs of top surface (a) LED grown on CSPSS, (b) CSPSS top surface after LLO process, (c) removed n-GaN, (d) after the LLO process

3.6 Surface treatment for higher extraction efficiency

Figure 11 (a) shows the SEM image of the top surface of the concavely patterned undoped-GaN layer after LLO process, transferred from the original CSPSS, with a uniform depth and size. The effective surface area of concavely patterned surface of VT-LED increased to about 50 % compared to that of planar surface VT-LED. The number of the based circle was calculated at about 10,000 based on the interval of ~1 μm and diameter of ~4 μm in 230 × 640 μm² device size (the device is covered with concave pattern about 75% in whole area). This concavely patterned structure reduces also the total internal reflection of the generated light in the active region of the LED and effectively scatters the light outward. For an additional efficiency improvement, concavely patterned surface was further roughened to nano-size dot by etching in 2 M KOH electrolyte at 85 ℃ for 7 min and 15 min, as shown in figure 11(b), (c), respectively. Although circular holes were generated from the initial CSPSS, the chemical etch revealed hexagonal-shaped holes. This is a result of the hexagonal crystal structure of wurtzite GaN. Especially, the concave region shows an open hexagonal inverted pyramid form which is defined by the six {10-1-1} planes.

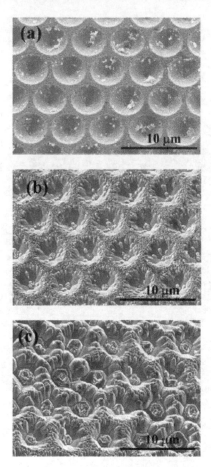

Fig. 11. SEM images of the top surface morphology removed GaN before(a) and after an etching time of 7 min(b) and 15 min(c) in 2M KOH at 85℃, respectively.

3.7 Chip separation using femtosecond laser scribing

Figure 12 shows schematics of a typical heat distribution in the substrate when the substrate is exposed to a focused long pulse and ultra fast short pulse laser beam. As the long pulse laser hits the substrate, the heat generated by the laser power greatly increases the temperature in the vicinity of the focused laser beam spot, locally melting the substrate, and further diffuses away into the material during the pulse duration, because the duration of laser pulse is longer than the heat diffusion time [Liu, et al. 1997; Lee, et al. 2010]. Therefore, the scribing speed is slower and fairly large amount of material droplet remains. However, when an ultrafast laser beam is used for the scribing, which is quite different from those of the long pulse laser beam. As the pulse widths decrease, the laser power intensity easily reaches hundreds of terawatts per square centimeter at the focused beam spot. The heat-affected volume is much smaller because the duration of the laser pulse is shorter than the heat diffusion time. Therefore, the scribing speed is faster and there is no damage.

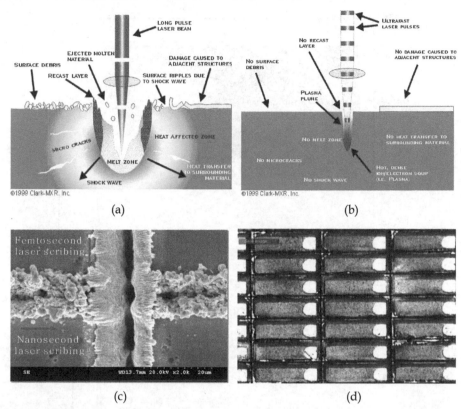

(a)   (b)

(c)   (d)

Fig. 12. Schematics of a typical heat distribution in the substrate when the substrate is exposed to a focused long pulse (a) and ultra fast short pulse (b) laser beam. (c) SEM images of the surface of Ni plating substrate after scribing with both nanosecond and femtosecond laser scribing in the same scribing time. (d) Chip separation by using femtosceond laser scribing process.

Figure 12 (c) shows SEM images of the surface of Ni plating substrate after scribing with both nanosecond and femtosecond laser scribing in the same scribing time. The scribing depth of femtosecond laser scribing showed a nearly sevenfold increase compared to that of nanosecond laser scribing. The fabricated VT-LEDs with 230×640 µm^2 size were separated by backside femtosceond laser scribing process as shown in Fig. 12 (d)

3.8 Characteristic of fabricated V-T LED

Figure 13 shows the current-voltage (I-V) characteristics of the fabricated VT-LED with the concavely patterned shape surface. Relatively low forward voltage (Vf =3.2 V) was observed at 20 mA operating current, with a very low reverse leakage current of -4 nA at -5 V. The insert of Fig. 13 shows the fabricated VT-LED with concave-patterned surface and the top view emission image under a driving current of 20 mA. Although it has a long length and the very small n-contact area for application of sideview LED in cellular phone, the current spreading reveals uniform over the total area. Due to a high electron mobility and conductivity

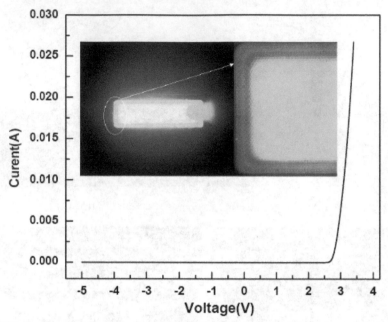

Fig. 13. I-V characteristics of the typical VT-LED. The top view emission images under a driving current of 20 mA in the inset.

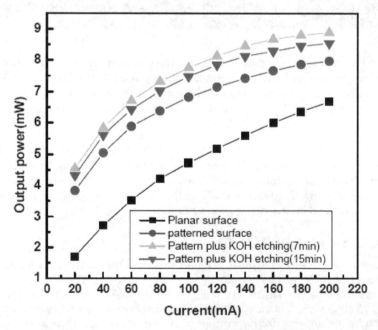

Fig. 14. Light output power of planar surface VT-LEDs and concavely patterned surface VT-LED before and after additional chemical etching in KOH, respectively.

of the n-GaN, the injected electron spreads well in the n-GaN layer without a semitransparent contact layer. Figure 14 shows the light output power as a function of injection current for these four VT-LEDs. The total output power of the devices were measured by using sideview PKG without phosphor to collect the light emitted in all directions from the LEDs. All luminous intensity (L-I) curves showed the slow-saturated characteristics up to 200 mA in relatively small chip size. Because of the higher thermal conductivity of nickel compared to sapphire, these devices are advantageous for high power operation. It should be noted that the output power at 20 mA was estimated to be 1.7, 3.8, 4.5, and 4.3 mW for a planar surface VT-LEDs, concavely patterned surface VT-LED before and after additional chemical etching in KOH, respectively. As compared with the output power for a planar surface VT-LED, the concavely patterned surface VT-LED using HPS plus KOH etching (7 min) resulted in an increase of output power by a factor of 2.7. This significant enhancement in output power could be attributed to the increase of the extraction efficiency, resulted from the increase in photon escaping probability due to enhanced light scattering at the concavely patterned and roughened surface with random nano-size dot.

4. Conclusion

In conclusion, The ELOG–like growth mode on the CSPSS effectively suppresses the probability of dislocation propagation into the cone region and hence greatly improves the crystal quality of the GaN films. The FWHM measured with asymmetric (102) rocking curves decreased from 338 arcsec for InGaN/GaN sample grown on CSS to 225 arcsec in the InGaN/GaN sample grown on CSPSS. To enhance light extraction efficiency, the use of a CSPSS is promising for the cost-effective production of better performing VT-LEDs. The output power of VT-LEDs with a concavely patterned surface is about 3.8 mW at a forward current of 20 mA, which is about 2.2 times higher than the devices with a flat surface. This improvement in the performance of VT-LEDs is attributed to the increase in the escaping probability of photons from the corrugated LED surface as well as the use of metal substrate.

5. Acknowledgment

The author thanks Dr. Jeong-Wook Lee and Mr. In-Suck Choi, Samsung LED Co., LTD., for their useful discussions.

6. References

Amano, H; Sawaki, N.; Akasaki, I. & Toyoda, Y. (1986). Metalorganic vapor phase epitaxial growth of a high quality GaN film using an AlN buffer layer, Appl. Phys. Lett., vol. 48, pp. 353.

Fujii, T.; Gao, Y.; Sharma, R.; Hu, E.L.; DenBaars, S. P. & Nakamura, S. (2004). Increase in the extraction efficiency of GaN-based light-emitting diodes via surface roughening, Appl. Phys. Lett., vol. 84, pp. 855–857.

Gil, B.(1998). Group III Nitride semiconductor Compounds," Clarendon, Oxford.

Grimmeiss, H & Koelmans, Z. H. (1959). Nature, vol. 14, pp. 264.

Huh, C.; Lee, K.S.; Kang E. J. & Park, S.J.(2003). Improved lightoutput and electrical performance of InGaN-based light-emitting diode by microroughening of the p-GaN surface, J. Appl. Phys., vol. 93, no. 11, pp. 9383–9385.

Johnson, W.C; Rasons, J. B & M. C. Crew, (1932). J. Phys. Chem., vol. 234, pp. 2651.

Juza, R; Hahn, H & Allg, Z. A. (1938). Metallic amides and metallic nitrides. V. Crystal structures of Cu3N, GaN and InN, Chem, vol. 234, pp. 282.

Khan, M. A.; Kuania, J. N.; Bhattarai, A.R. & Olsen, D.T. (1993). Metal semiconductor field effect transistor based on single crystal GaN, Appl. Phys. Lett., vol. 62, pp. 1786.

Kittel, C. (1995). Introduction to Solid state physics, 7th Edition, Wiley.

Lee, J. H.; Oh, J.T.; Choi,S.B.;Kim, Y.C.; Cho, H.I. & Lee, J.H.(2008). Enhancement of InGaN-based vertical LED with concavely patterned surface using patterned sapphire substrate," IEEE Photon. Technol. Lett., vol. 20, no. 5, pp. 345–347.

Lee, J. H.; Oh, J.T.; Kim, Y.C & Lee, J.H.(2008). Stress reduction and enhanced extraction efficiency of GaN-based LED grown on cone shape-patterned sapphire, IEEE Photon. Technol. Lett., vol. 20, no. 18, pp. 1563–1565.

Lee, J.H.; Kim, N.S.; Hong, S.S.& Lee, J.H. (2010).Enhanced extraction efficiency of InGaN-based light-emitting diodes using 100 kHz femtosecond-laser-scribing technology, IEEE Electron Device Lett., vol. 31, pp. 213-215.

Lee, J.H.; Lee, D.Y.; Oh, B.W. & Lee, J.H.(2010). Comparison of InGaN-based LEDs grown on conventional sapphire and cone-shapepatterned sapphire substrate, IEEE Trans. Electron Devices, vol. 57, no. 1, pp. 157–163.

Lee, J.H.; Oh, J.T.; Park, J.S.; Kim,J. W.; Kim, Y.C.; Lee, J.W. & Cho, H.K.(2006). Improvement of luminous intensity of InGaN light emitting diodes grown on hemispherical patterned sapphire, Phys. Stat. Sol. (c), vol. 3, pp. 2169–2173.

Levinshtein, M; Rumyantsev, S. & M. Shur, M. (2001). Properties of Advanced Semiconductor Materials, Wiley, New York.

Liu, X.; Du, D. & Mourou, G.(1997). Laser ablation and micromachining with ultrashort laser pulses, IEEE J. Quantum Electron, vol. 33, pp. 1706-1716.; Http://www.cmxr.com.

Look, D.C. & Sizelove, J.R.(2001). Predicted maximum mobility in bulk GaN, Appl. Phys. Lett., vol. 79, pp. 1133.

Maruskam, H. P. & Tietjen, J. J. (1969). The preparation and properties of vapor deposited single crystal line GaN, Appl. Phys. Lett., vol. 15, pp. 327.

Nakamura, S. & G. Fason, G.(1997). The Blue Laser Diode, Sringer, Berlin.

Nakamura, S.(1991). GaN Growth Using GaN Buffer Layer, Jpn. J. Appl. Phys., vol. 30, pp. L1705, 1991.

Nakamura, S.; Senoh, M.; Iwasa, N. & Nagahama, S. (1995). High brightness InGaN blue, green and yellow light-emitting-diodes with quantum well structure, Jpn. J. Appl. Phys., vol. 34, no. 7A, pp. L797–L799.

Nakamura, S.; Senoh, M.; Nagahama, S.; Iwasa, N.; Yamada, T.; Mat-sushita, T.; Sugimoto, Y. & Kiyoku, H. (1997). High-power, long-lifetime InGaN multi-quantum-well-structure laser diodes, Jpn. J. Appl. Phys., vol. 36, no. 8B, pp. L1059–L1061.

Nakamura, S.; Senoh, M.; Nagahoma, S.; Iwasa, N.; Yamada, T.; Matsushita, T.; Kiyoku, H. & Sugimoto, S. (1996). InGaN multi-quantum-well structure laser diodes grown on MgAl2O4 substrates, Appl. Phys. Lett., vol. 68, pp. 2105.

Ng, H.M.; Doppalapudi, D.; Moustakas, T.D.; Weimann, N.G. & Eastman, L.F.(1998). The role of dislocation scattering in n-type GaN films, Appl. Phys. Lett., vol. 73, pp. 821.

Pankove, J. I. (1997). GaN and Related Materials, Gordon and Breach, New York.

Pankove, J. I.; Chang, S. S.; Lee, H.C.; Molanar, R.; Moustakas, T.D. & Zeghbroeck, B., (1994). High-temperature GaN/SiC heterojunction bipolar transistorwith high gain , Tech. Dig. Int. Electron Devices Meet., vol. 94, pp.389.

Pankove, J. I; Miller, E. A. & Berkeyheiser, J. E. (1972). GaN blue light-emitting diodes, J. Limin. vol. 5, pp. 84.

Ridley, B.K.; Schaff, W.J. & Eastman, L.F. (2004). Hot-phonon-induced velocity saturation in GaN, Appl. Phys. Lett., vol. 96, pp. 1499.

Sakai, A.; Sunakawa, H. & Usui, A. (1997). Defect structure in selectively grown GaN films with low threading dislocation density, Appl. Phys.Lett., vol. 71, no. 16, pp. 2259–2261.

Simeonov, D; Fetin, E; Carlin, J.F.; Ilegems, M. And N. Grandjean, N.(2006). Stranski–Krastanov GaN/AlN quantum dots grown by metal organic vapor phase epitaxy, J. Appl. Phys., vol. 99, pp. 083509.

Smorchkova, I.P.; Keller, S.; Heikman, S.; Elsass, C.R.;Heying, B.; Fini, P.; Speck, J.S. & Mishra, U.K.(2000).Two-dimensional electron-gas AlN/GaN heterostructures with extremely thin AlN barriers, Appl. Phys. Lett., vol. 77, no. 24, pp. 3998–4000.

Sugiura, L.; Itaya, K.; Nishio, J.; Fujimoto, H. & Kokubun, Y.(1997). Effects of thermal treatment of low-temperature GaN buffer layers on the quality of subsequent GaN layers, J. Appl. Phys., vol. 82, pp. 4877–4882.

Sumiya, M,; Ogusu, N.; Yotsuda, Y.; Itoh, M.; Fuke, S.; Nakamura,T.; Mochizuki, S.; Sano, T.; Kamiyama, S.; Amano, H. & Akasaki, I.(2003). Systematic analysis and control of low-temperature GaN buffer layers on sapphire substrates, J. Appl. Phys., vol. 93, pp. 1311–1319.

Tadatomo, K.; Okagawa, H.; Ohuchi, Y.; Tsunekawa, T.; Imada, Y.; Kato, M.& Taguchi, T. (2001). High output power InGaN ultraviolet lightemitting diodes fabricated on patterned substrates using metalorganic vapor phase epitaxy, Jpn. J. Appl. Phys., vol. 40, pp. L583–L585.

Wang, S.J.; Uang, K. M.; Chen, S.L.; Yang, Y. C; Chang, S. C.; Chen, T. M. & Chen, C.H. (2005). Use of patterned laser liftoff process and electroplating nickel layer for the fabrication of vertical-structured GaN-based light-emitting diodes," Appl. Phys. Lett., vol. 87, pp. 011111–011113, 2005.

Wang, W. K.; Huang, S. Y.; Huang, S.H.; Wen, K.S.; Wuu, D.S. & Horng, R.H. (2006).Fabrication and efficiency improvement of micropillar InGaN/Cu light-emitting diodes with vertical electrodes," Appl. Phys. Lett., vol. 88, pp. 181113–181115.

Wraback, M.; Shen, H.; Carrano, J.C.; Li, T.; Campbell, J.C.; Schurman, M.J. & Ferguson, I.T.(2000). Time-resolved electroabsorption measurement of the electron velocity-field characteristic in GaN, Appl. Phys. Lett., vol. 79, pp. 1155.

Wuu, D.S.; Wang, W.K.; Wen, K.S.;Huang, S.C.; Lin, S. H.;Horng, R.H.;Yu,Y.S. & Pan, M.H. (2006). Fabrication of pyramidal patterned sapphire substrates for high-efficiency InGaN-based light emitting diodes, J. Electrochem. Soc., vol. 153, pp. G765–G770.

Yamada, M.; Mitani, T.; Narukawa, Y.; Shioji,S.; Niki, I.; Sonobe, S.; Deguchi, K.; Sano, M. & Mukai, T.(2002). nGaN-based near-ultraviolet and blue-light-emitting diodes with high external quantum efficiency using a patterned sapphire substrate and a mesh electrode, Jpn. J. Appl. Phys., vol. 41, pp. L1431–1433.

Yoshida, S; Misawa, S. & Gonda, S. (1983). Improvements on the electrical and luminescent properties of reactive molecular beam epitaxially grown GaN films by using AlN coated sapphire substrates, Appl. Phys. Lett., vol. 42, pp. 427.

Youn, D.H.; Lee, J.H.; Kumar, V.; Lee, K.S.; Lee, J.H. & Adesida, I.(2004) The effects of isoelectronic Al doping and process optimization for the fabrication of high-power AlGaN-GaN HEMTs, IEEE Trans. Electron Devices, vol. 51, no. 5, pp. 785–789.

Zheleva,T.S.; Nam, O.H.; Bremser, M.D. and R. F. Davis, R. F.(1997). Dislocation density reduction via lateral epitaxy in selectively grown GaN structures, Appl. Phys. Lett., vol. 71, no. 17, pp. 2472–2474.

Part 2

Environmental Issue

5

Resistant Fungal Biodiversity of Electroplating Effluent and Their Metal Tolerance Index

Arifa Tahir

Environmental Science Department, LCWU Lahore,
Pakistan

1. Introduction

Discharge of heavy metals in the aquatic system has become a global phenomenon due to their carcinogenic and mutagenic nature (Mahiva *et al.*, 2008). electroplating industry in Pakistan is contributing its major part in deteriorating the country environment at massive scale with the accumulation of heavy metals in aqueous environment. The chemically polluted water has seriously damaged the ecology of surface and ground water, which eventually impart serious consequences on agriculture due to contamination of crops grown in a polluted area .

The introduction of heavy metal compounds into the environment generally induces morphological and physiological changes in the microbial communities (Vadkertiova and Slavikova, 2006), hence exerting a selective pressure on the microbiota (Verma et al., 2001). Strains isolated from contaminated sites have an excellent ability of removing significant quantities of metals from effluents (Malik, 2004). Conventional processes such as chemical precipitation; ion exchange and reverse osmosis are uneconomical and inefficient for treating effluents (Gupta et al., 2000; Pagnanelli et al., 2000; Gavrileca, 2004; Malik, 2004).Biosorption, using bacteria, fungi, yeast and algae, is regarded as a cost-effective biotechnology for the treatment of wastewaters containing heavy metals. Among the promising biosorbents for heavy metal removal which have been researched during the past decades, fungi has received increasing attention due to their higher ability to remove high concentrations of heavy metals than yeast, bacteria and and algae ((Gavrilesca, 2004; Baldrian, 2003, Zafar et al., 2007). They can adapt and grow under high metal concentrations (Anand et al., 2006). They offer the advantage of having cell wall material which shows excellent metal-binding properties (Gupta et al., 2000). Fungi are known to tolerate metals by several mechanisms including valence transformation, extra and intracellular precipitation and active uptake (Malik, 2004). Considering the above mechanisms of metal resistance in fungi, we have studied filamentous fungi isolated from electroplating industrial effluent, a polluted environment to assess their metal tolerance and metal removal potential from aqueous solution. High affinity, rapid rate of metal uptake and maximum loading capacity are important factors for the selection of a biosorbent. Therefore, there is an increased interest in the identification of some new and better biosorbents that show promising uptake of metallic ions (Akhtara *et al.*, 2007). It is expected that screening of metal tolerant fungi may provide strains with improved metal accumulation. The introduction of heavy metal compounds into the environment generally induces morphological and physiological changes in the microbial communities.The aim of the present study was to study Cu resistant fungal biodiversity and their metal tolerance index from electroplating effluent. The study was also aimed to analyse functional group responsible for Cu removal. There is huge literature on the biosorption of heavy metals from electroplating effluent but there was no attention on the *Gliocladium* sp. as potential biosorbent. In this present work *Gliocladium viride* was selected after screening of 50 fungal isolates. Potential fungal strain was selected on its comparative biosorption capacity.

To the best of our knowledge this work is the first report on biosorption of Cu by *Gliocladium viride*. This non-pathogenic fungus belongs to Ascomycota division. It is well known for the production of lytic enzymes (cellulases and chitinases) and antibiotics. It also acts as a biocontrol agent against plant pathogens (Druzhinina *et al.*, 2005). The effluent samples were characterized for its physicochemical parameters.A total of fifty filamentous fungal strains representing five genera; *Gliocladium, Penicillium, Aspergillus, Rhizopus* and *Mucor* species, were isolated and screened for their Cu removal efficiencies from electroplating tanning effluent. The experimental results showed that *Gliocladium* sp. was the best Cu resistant fungus among all fungal species isolated from the effluent. These strains were exposed to Cu metal ions up to 3mM to study tolerance of isolated fungal.The degree of tolerance was measured from the growth rate in the presence of Cu. Whole mycelium and cell wall component were analyzed for Cu biosorption. Cell wall component was found to be responsible for Cu biosorption. Amino groups were found to be abundant in the cell wall of *Gliocladium viride* ZIC_{2063} as determined by infrared spectroscopy. These examinations indicated the involvement of amines in metal uptake.

2. Matherial and methods

2.1 Isolation and scerening of fungi for cutolerance

The PDA (Potato Dextrose Agar) medium was used for culturing fungi from tanning effluent. One ml of serially diluted (1000-fold) effluent was spread on the PDA agar plate. The inoculated plates were incubated at 30 ^{0}C for 72 h. The morphologically distinct colonies were selected and screened for their Cutolerance. A small piece of mycelium (3 mm^2) was inoculated on PDA plates supplemented with 3mM copper sulphate.The plates were incubated for seven days at 25 ^{o}C. The control (without metal solution) was also run in parallel. Reduction of radial growth rate was used as an index for metal tolerance. To compare the heavy metal tolerance of each isolate, a parallel index of tolerance (T.I) was calculated as a percentage value from the ratio:

$$T.\, I. = \frac{\text{Radial growth rate in metal treatment}}{\text{Radial growth rate in control}}$$

The isolates exhibiting better growth after incubation were considered as tolerant to the metal.

2.2 Biosorption studies

All fungal strains were further screened to check their Cu removal efficiencies. Potential of fungal strains for Cu removal was evaluated in batch studies. For the preparation of fungal pellets, 7-days-old spore inoculum (10 %) of each strain was inoculated into sterile 50 ml Potato Dextrose Broth in a 250 ml conical flask and incubated for 96 hour at 30 °C in an orbital shaker (122 rpm). Mycelial pellets were filtered through cheesecloth. These fungal pellets (2.0 g wet weight) were added into 10% diluted effluent (pH 3.0) in 250 ml conical flask and incubated at 30 °C for 24 h at 122 rpm. The control was also run in parallel containing diluted effluent without fungal pellet. After 24 h of contact time, samples were centrifuged, filtered and supernatant was checked for residual Cu metal ions concentration. The potential strain with maximum Cu removal rate was selected for further work.

2.3 Identification

All isolates were identified on the basis of their morphological characteristics (colonial morphology, color, texture, shape, diameter and appearance of colony,) and microscopic characteristics (septation of mycelium, shape, diameter and texture of conidia).

2.4 Isolation of cell wall

The cell wall fraction of isolated fungal strains was separated by the method of Baik *et al*, 2002. Fungal mycelium (20.0 g) was homogenized for 15 min in a blender (National, Japan). Homogenized mycelium was washed with water and centrifuged for 30 min at 8000 rpm. The pellet was mixed with 250mL of mixture of chloroform and methanol (1:1). Then it was washed twice with acetone, and once with ethanol. The cell wall yield was 254.65 mg, 287.02 mg, 246.75 mg, 310.8 mg and 280 mg from 1.0 g fungal biomass of *Penicillium* sp., *Rhizopus sp*, *Mucor* sp., *Gliocladium viride* ZIC_{2063} and *Aspergillus* sp. The biosorption potential of cell wall component was determined. The cell wall (2.0 g) was suspended in 50 ml effluent in 250 ml conical flask and incubated at 30 °C for 24 h at 122 rpm.

2.5 FTIR spectrometer analysis

To characterize the Infrared spectra of fungal sp. Fourier Transform Infrared Spectrometer M 2000 series (MIDAC Corporation, Irvine California) was used.

2.6 Analytical Method for copper (VI)

The heavy metal concentration was determined by the use of Polarized Zeeman Atomic Absorption Spectrophotometer Z-5000 (Perkin Elmer Analyst 300 Hitachi, Japan). Determination of copper was done by using its specific lamp and at a specific wavelength. Samples (almost 5 ml each) of first 1h biosorption taken at predetermined interval were centrifuged and filtered. The supernatant were analyzed for Cu concentration.

The amount of metal bound by the biosorbents was calculated as follows.

$$Q = V (C_i - C_f) / m$$

Where,

Q = Metal uptake (mg metal per g of biosorbent),

V = Liquid sample volume (ml),

Ci = Initial concentration of the metal in the solution (mg/L),

Cf = Final concentration of the metal in the solution (mg/L) and

m = Amount of the added biosorbent on dry basis (mg).

3. Results and discussion

A total of fifty fungal strains were isolated from tanning effluent. Among these 11 isolates were of *Penicillium* sp., 7 of *Aspergillus* sp., 17 of *Gliocladium* sp., 6 of *Rhizopus* sp. and 9 isolates of *Mucor* sp. Pollution of water by heavy metals may lead to decrease in microbial diversity and enhanced the growth of resistant species (Ezzouhri *et al.*, 2009). The data of table 1 also showed the metal tolerance index of the isolates. *Mucor* sp. were found to be very much sensitive to Cu which shows that Cu is highly toxic to its growth while thick mycelium of *Gliocladium viride* was observed in the presence of Cu than in control.

Among all fungal species *Gliocladium viride* showed higher efficiency to remove Cu. The removal efficiencies of fungal biosorbents for Cudecreased in the order: *Gliocladium* sp. > *Penicillium* sp. > *Aspergillus* sp. > *Rhizopus* sp. > *Mucor* sp. (Table.2). The statistical analysis of the data showed high significant level (89.48) for Cuuptake by *Gliocladium* viride ZIC_{2063} as compared to other isolates. The Curemoval efficiency of *Gliocladium viride* ZIC_{2063} (91.97 %) was much greater than other biosorbents reported by other researchers such as *Rhizopus nigricans* 80 % (Bai and Abraham, 2001), *Aspergillus niger* 83 % (Mala *et al.*, 2006), *Rhizopus oryzae* 23 % (Park and Park, 2005) and *Sargassum* sp. 60 % (Cossich *et al.*, 2002).

FTIR spectrum of all isolated fungi was also studied. FTIR spectrum of *Gliocladium virid* ZIC_{2063} has absorption peaks at 3922, 3467, 2393, 2354, 2055, 1634, 1072 and 517 cm[-1] frequency level (data not shown). Absorption peaks of *Gliocladium viride* ZIC_{2063} indicates the presence of hydroxyl groups, Primary and secondary amines, amides stretching, Imines, oximes, C-OH stretching and C-N-C bonding are present on its cell wal (Fig 1)l. Biosorption process occurred at amino, hydroxyl and carboxyl groups present in cellulose and chitin of fungal cell wall (Ozsoy *et al.*, 2008). FTIR spectrum analysis of *Gliocladium viride* ZIC_{2063} showed that amine and its derivatives are the most common functional groups attached on its cell surface. Our results support the finding of other workers. According to Bai and Abraham, 2002 amino groups of fungal cell wall are mainly responsible for metal biosorption. metal ions did not bind to negatively charged functional groups such as carboxylate, phosphate and sulphate. Only positively charged amines (protonated at low pH) are responsible for binding of metal ions (Bayramoglu *et al.*, 2005).A single experiment was conducted to test the metal uptake capacity of whole mycelium and cell wall component (Table.2). Highest metal uptake (1230.2 mg) was found in cell wall component of *Gliocladium viride* ZIC2063. The cell wall components of other fungal isolates also gave greater metal uptake capacity than whole mycelium. It is in contrasted with other workers who reported greater metal uptake by whole mycelium than cell wall (Baik *et al*, 2002).

No.	Isolates No.	Removal rate (%)	Cu tolerance index
1	Penicillium TEI 1220	73.75	0.3
2	Penicillium TEI 1221	84.98	0.25
3	Penicillium TEI 1222	77.15	0.19
4	Penicillium TEI 1223	70.71	0.29
5	Penicillium TEI 1224	69.07	0.3
6	Penicillium TEI 1225	82.95	0.5
7	Penicillium TEI 1226	83.39	0.18
8	Penicillium TEI 1227	74.70	0.2
9	Penicillium TEI 1228	70.73	0.27
10	Penicillium TEI 1229	72.42	0.26
11	Penicillium TEI 1230	72.62	0.2
12	*Aspergillus* .ESGO2001	60.59	0.16
13	*Aspergillus* ESGO 2002	62.15	0.2
14	*Aspergillus* ESGO 2003	66.72	0.5
15	*Aspergillus* ESGO 2004	65.25	0.35
16	*Aspergillus* ESGO 2005	67.36	0.49
17	*Aspergillus* ESGO 2006	62.61	0.35
18	*Aspergillus* ESGO 2007	59.50	0.5
19	*Gliocladium* ZIC 2060	86.65	1.3
20	*Gliocladium* ZIC2061	85.71	1.39
21	*Gliocladium* ZIC 2062	89.75	1.41
22	*Gliocladium* ZIC 2063	96.98	1.42
23	*Gliocladium* ZIC 2064	79.74	1.38
24	*Gliocladium* ZIC 2065	91.60	1.32
25	*Gliocladium* ZIC 2066	87.49	1.4
26	*Gliocladium* ZIC 2067	83.03	1.38
27	*Gliocladium* ZIC 2068	79.61	1.31
28	*Gliocladium* ZIC 2069	87.69	1.33
29	*Gliocladium* ZIC 2070	83.29	1.36
30	*Gliocladium* ZIC 2071	84.94	1.39
31	*Gliocladium* ZIC 2072	92.43	1.4
32	*Gliocladium* ZIC 2073	90.45	1.32
33	*Gliocladium* ZIC 2074	82.84	1.35
34	*Gliocladium* ZIC 2075	91.48	1.37
35	*Gliocladium* ZIC 2076	86.60	1.35
36	*Rhizopus* SID 090	61.38	0.19
37	*Rhizopus* SSID 091	57.70	0.2
38	*Rhizopus* SSID 092	62.85	0.21

No.	Isolates No.	Removal rate (%)	Cu tolerance index
39	*Rhizopus* SSID 093	55.16	0.22
40	*Rhizopus* SSID 094	59.11	0.09
41	*Rhizopus* SSID 095	55.87	0.21
42	*Mucor* .CENTA 001	47.84	0.06
43	*Mucor* CENTA 002	51.01	0.04
44	*Mucor* CENTA 003	48.46	0.056
45	*Mucor* CENTA 004	50.36	0.06
46	*Mucor* CENTA 005	49.08	0.05
47	*Mucor* CENTA 006	45.50	0.09
48	*Mucor* CENTA 007	49.35	0.08
49	*Mucor* CENTA 008	50.62	0.1
50	*Mucor* CENTA 009	54.86	0.09

Biosorption Conditions: Incubation time 24 h, Temperature 30 °C, pH 3.0, Biosorbent 2.0 g (wet weight), agitation 122 rpm, Volume of reaction mixture 50 ml.

Table 1. Biosorption potiental of fungal isolates for their Cu removal efficiencies

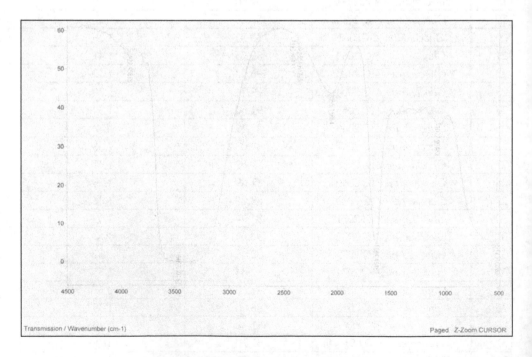

Fig. 1. FTIR spectrum of *Gliocladium viride* ZIC$_{2063}$.

Sr.No	Organism	Biosorbent material	Cu binding capacity (mg/g)
1	Penicillum sp.	Whole mycelium	254.7
		Cell wall	964.5
2	Aspergillus sp.	Whole mycelium	126.7
		Cell wall	894.6
3	Gliocladium sp.	Whole mycelium	474.5
		Cell wall	1230.2
4	Rhizopus sp.	Whole mycelium	98.8
		Cell wall	726.2
5	Mucor sp.	Whole mycelium	94.6
		Cell wall	541.8

Table 2. Cubinding capacity of cell components of different fungal isolates.

4. Conclusion

In this study, copper resistant fungi were isolated from heavy metal contaminated environments, and the applicability of their heavy metal removal from industrial wastewater was evaluated at a laboratory scale. The heavy metal removal was determined for each isolate. *Gliocladium viride* was found to be highly copper tolerant fungus and exhibited thick growth than other fungal sp ecies. It appears that *Gliocladium* species has greater Cu removal efficiency (96.98 %) than other fungal species. Our findings also indicate direct relationship between level of metal resistance and biosorption capacity. Further investigations are underway to optimize the conditions for Cu removal from industrial effluent. *Gliocladium viride* ZIC_{2063} can be exploited as potential biosorbent for Cu from electroplating effluent.

5. References

Akhtara K, Akhtarb MW and Khalid AM (2007). Removal and recovery of uranium from aqueous solutions by *Trichoderma harzianum*. *Water Research*, **41**:1366 – 1378.

Anand P, Isar J, Saran S, Saxena RK (2006). Bioaccumulation of copper by Trichoderma viride. Bioresour. Technol. 97: 1018-1025.

Bai RS and Abraham TE(2002). Studies on enhancement of Cr (VI) biosorption by chemically modified biomass of Rhizopus nigricans.Water Res. 36: 1224–1236.

Baik WY, Bae JH, Cho KM and Hartmeier W.(2002) Biosorption of heavy metals using whole mold mycelia and parts thereof.Bioresource Technology. 81, :167-170

Baldrian P, Gabriel J (2002). Intraspecific variability in growth response to cadmium of the wood-rotting fungus Piptoporus betulinus .Mycologia. 94: 428-436.

Bayramoglu G, Celik G, Yalcın E, Yılmaz M. and Arıca MY. (2005).Modification of surface properties of Lentinus sajor-caju mycelia by physical and chemical methods: evaluation of their Cr6+ removal efficiencie from aqueous medium. Journal of Hazardous Materials, 119: 219–229.

Bennet JW, Wunch KG and Fatson BD (2002). Use of Fungi Biodegradation,Manual of Environmental Microbiology Second Edition Editor in Chief Christon J. Hurst ASM Press Washington, D.C.

Cossich ES, Tavares CR. G. and Ravagnani TMK. (2002). Biosorption of chromium (III) by Sargassum sp. Biomass. Electronic Journal of Biotechnology, 5: ISSN 0717-3458.

Druzhinina IS, Kopchinskiy AG, Komoj M, Bissett J, Szakacs G And Kubicek CP (2005). Fungal Genetics and Biology, 42: 813–828.

Ezzouhri1,L. Castro,E. Moya,M. Espinola, F. and Lairini,K. (2009).Heavy metal tolerance of filamentous fungi isolated from polluted sites in Tangier, Morocco African Journal of Microbiology Research . 3 (2) pp. 035-04

Gavrilesca M (2004). Removal of heavy metals from the environment by biosorption. Eng. Life Sci. 4(3): 219-232.

Gupta VK, Shrivastava AK and Neeraj J. (2001). Biosorption of chromium (VI) from aqueous solutions by green algae Spirogyra species.water Research, 17: 4079-4085.

Gupta R, Ahuja P, Khan S, Saxena RK, Mohapatra H (2000). Microbial biosorbents:Meeting challenges of heavy metal pollution in aqueous solutions.Curr. Sci. 78(8): 967-973.

Mahvi AH., Gholami F, Nazmara S, 2008. Cadmium adsorption from wastewater by almas leaves and their ash. Euro. Sci. Res., 23: 19

Mala JGS, Nair BU and Puvanakrishnan R. (2006). Accumulation and biosorption of chromium by Aspergillus niger MTCC 2594. J. Gen. Appl. Microbiol., 52: 179–186.

Malik A (2004). Metal bioremediation through growing cells. Environ. Int. 30: 261-278.

Ozsoy HD, Kumbur H, Saha B, van Leeuwen JH (2008). Use of Rhizopus oligosporus produced from food process wastewater as a biosorbent for Cu(II) ions removal from the aqueous solutions. Bioresour. Technol. 99: 4943-4948.

Pagnanelli F, Petrangeli MP, Toro L, Trifoni M, Veglio F (2000).Biosorption of metal ions on Anthrobacter sp.: Biomass characterization and biosorption modelling. Environ. Sci. Technol.34(13): 2773-2778.

Park D, Yun YS and Park JM (2005). Use of dead fungal biomass for the detoxification of hexavalent chromium: screening and kinetics. Process Biochemistry, 40:2559–2565.

Vadkertiova R, Slavikova E (2006). Metal tolerance of yeasts isolated from water, soil and plant environments. J. Basic Microbiol. 46: 145-152.

Verma T, Srinath T, Gadpayle RU, Ramteke PW, Hans RK, Garg SK (2001). Chromate tolerant bacteria isolated from tannery effluent.Bioresour. Technol. 78: 31-35.

Zafar S, Aqil F and Ahmad, I. (2007). Metal tolerance and biosorption potential filamentous fungi isolated from metal contaminated agricultural soil. Bioresource Technology, 98: 2557–2561.

6

Biomass Adsorbent for Removal of Toxic Metal Ions From Electroplating Industry Wastewater

Ronaldo Ferreira do. Nascimento[1], Francisco Wagner de Sousa[1],
Vicente Oliveira Sousa Neto[2], Pierre Basílio Almeida Fechine[1],
Raimundo Nonato Pereira Teixeira[3], Paulo de Tarso C. Freire[1]
and Marcos Antônio Araujo-Silva[1]
[1]Universidade Federal do Ceara (UFC),
[2]Universidade Estadual do Ceara (UECE-CECITEC),
[3]Universidade Regional do Cariri (URCA),
[4]Instituto Federal de Educação, Ciência e Tecnologia (IFCE-Crateús)
Brazil

1. Introduction

The contamination of water bodies by heavy metals has been the subject of many studies worldwide. Several alternatives have been proposed in order to minimize the harmful effects that the disposal of these metals can cause to the environment.

A major challenge for the electroplating industry is finding solutions that equate to positive environmental and economic aspects regarding the treatment of their effluent. The adsorption process has been widely studied in order to solve the problem of many industries regarding the disposal of their effluent. Systems such as ion exchange resins, electrochemical process, chemical precipitation and activated carbon have been widely used in the processes of waste water purification. Such processes could have reduced its costs from the use of alternative materials for low cost. This has led many industries worldwide to invest in research aimed at obtaining cheap and plentiful materials that may have the same or greater capacity to remove pollutants from water bodies.

The biosorption has been presented as a promising alternative to solve the problem of contamination by heavy metals with low environmental and economic impacts.

In this chapter we introduce biosorption e process of heavy metals and discuss the implications of the technologies used in the electroplating industry to remove its contaminants.

1.1 Biosorbents

Materials of natural origin are generally used in biosorption studies (such as seaweed, biological depuration plant sludge, agricultural and industrial wastes) are inexhaustible, low-cost and non-hazardous materials, which are specifically selective for different contaminants and easily disposed by incineration.

The biosorbent terms refers to material derived from microbial biomass, seaweed or plants that exhibit adsorptive properties. Biosorption is the accumulation of metals by biological materials without active uptake. This process may include ion exchange, coordination, complexation, chelation, adsorption and microprecipitation (Duncan et al., 1994) This biomass must be subjected to pre-treatment to obtain a better operating performance in pollutant removal using adsorption method (Naja et al., 2010).

An interesting feature is that the biosorbents are widely found in nature and they have low cost. In some cases they are agriculture waste such as corn cob (Shen & Duvnjak, 2005), coconut shell (Sousa, 2007), orange pulp (Almeida, 2005), peat (Gupta et al 2009) and sawdust (Yasemin & Zaki, 2007)

1.2 Ions metal biosorption

The process is based on the interaction of ions at the interface biomass / aqueous medium. The sorbent can be either a particulate material as a compact material. The separation can be performed in packed columns, fluidized beds or in the form of discs to be used in the filtration process. This configuration allows for regeneration and reuse of the adsorbent and its proper disposal (Vargas et al, 1995). The discovery and development of the biosorption process has supported the basis for an new technology in this field. Several authors have been devoted to the study and applicability of this new kind of technology (White, 1995; Volesky, 1990).

Agricultural waste is one of the rich sources of low-cost adsorbents besides industrial by-product and natural material. Due to its abundant availability agricultural waste such as peanut husk, rice husk, coconut shell, wheat bran and sawdust offer little economic value and, moreover, create serious disposal problems (Igwe & Abia,2007).

Biosorption is the capability of active sites on the surface of biomaterials to bind and concentrate heavy metals from even the most dilute aqueous solutions. Biosorption can be used for the treatment of wastewater with low heavy metal concentration as an inexpensive, simple and effective alternative to conventional methods. The process of metal ion binding is comprised of many physico-chemical processes like ion exchange, complexation, microprecipitation, and electrostatic interactions. Biosorbents for the removal of metals mainly come under the following categories: bacteria, fungi, algae, industrial wastes, agricultural wastes and other polysaccharide materials. In general, all types of biomaterials have shown good biosorption capacities towards all types of metal ions (Febrianto et al., 2009)

The biosorption results from electrostatic interactions and/or of the formation of complexes between metal ions and functional groups present on the surface of biosorbent. So many studies have been conducted to assess the potential for removing heavy metals from various biological materials (Hayashi , 2001). We can cite the seaweeds and their derivatives (Luo et al, 2006), chitosan (Ngah et al, 2011), lignin (Guo et al, 2008), wide variety of bacteria and fungi (Watanabe et al, 2003), agricultural residues (Sousa, 2007), among others. According to Volesky, (1990) biosorbent must have certain physical characteristics (surface area, porosity, grain size etc.) to get good ability to adsorption and to be used in an adsorption process.

1.3 Adsorbent used for removing pollutants

Adsorption process is essentially a surface phenomenon. Adsorbent having a good adsorptive capacity implies that it should present a large specific surface area. The adsorptive properties depend on the distribution of size pore and the nature of the solid surface. The adsorbents most commonly used on an industrial scale are the activated carbon, silica gel, activated alumina and molecular sieves (Lopez & Gutarra, 2000) (Yasemin & Zaki, 2007). There is no doubt that the charcoal has become the most widely used solid, worldwide, as an adsorbent to remove pollutants in wastewater. According to its own characteristics, such as high porosity, chemical structure and high surface area, the activated carbon has an excellent ability to remove substances on its surface (Babel & Kurniawan, 2003). Added to this, the chemical structure of this material allows surface modifications by chemical or physical treatments, allowing an increase in adsorption capacity of this material.

On the other hand, it is also widespread agreement that even with all these features, the charcoal has some serious drawbacks to adsorptive processes. Some of them should be, for example, the fact that this material is not selective, and its market value is relatively high. Also, the reactivation of this material is not an easy process; cleaning the surface for subsequent applications is a one very expensive process. Moreover, the methods of recovery of coal, both the heat treatment as the use of chemicals, cannot regenerate the material with the same initial characteristics, leading to losses in their adsorption capacity. For these reasons, special attention has been focused on several other adsorbent materials. In particular, some natural materials, such as polysaccharides, clays, biomass, etc., that can remove pollutants from contaminated water at low cost of procurement has been widely researched around the world (Kumar, 2000; Crini, 2005, Crini, 2006). In fact, the cost of obtaining and regenerative capacity of such solid materials are important parameters when compared adsorbents materials.

№	pH	Adsorbent	Metal	Adsorption capacity $q_m(mg.g^{-1})$	Reference
1	-	Pomelo peel	Cd(II)	21.83	(Saikaew et al, 2009)
2	6	Acid-treated coconut shell carbon	Zn(II)	60.41	(Amuda et al, 2007)
3	5	Caladium bicolor	Cd(II)	42.19	(Jnr et al, 2005)
4	-	H_3PO_4 reated rice bran	Ni(II)	102	(Zafar et al,2007)
5	6.6-6.8	Rice husk modified	Cd(II)	20.24	(Kumar and Bandyopadhyay, 2006)
6	5.3	Gelidium	Zn(II) Cr(III)	13 18	(Vilar et al, 2007)
7	5.3	Algal waste	Zn(II) Cr(III)	7.1 11.8	(Vilar et al, 2007)
8	5.0	Valonia tannin resin	Cu(II)	44.24	(Sengil et al, 2009)
9	5.5	Lignin	Cd(II)	25.4	(Guo, 2008)
10	5.3	Modified wood	Cu(II) Pb(II)	23.7 82.6	(Low et al,2004)

Table 1. Maximum adsorption capacity of some adsorbent system given in literature

1.4 Modification of adsorbent

In general, raw lignocellulosic biosorbents were modified by various methods to increase their adsorption capacities because metal ion binding by lignocellulosic biosorbents is believed to take place through chemical functional groups such as carboxyl, amino, or phenolics. More recently, great effort has been contributed to develop new adsorbents and improve existing adsorbents. Many investigators have studied the feasibility of using low-cost agro-based waste materials (Hashem, 2006; Hashem et al.,2006; Hashem et al. ,2006; Hashem et al., 2006; Hashem et al., 2006, Abdel-Halim et al., 2006; Hashem et al., 2006)

Agricultural by-products usually are composed of lignin and cellulose as major constituents and may also include other polar functional groups of lignin, which includes alcohols, aldehydes, ketones, carboxylic, phenolic and ether groups. These groups have the ability to some extent to bind heavy metals by donation of an electron pair from these groups to form complexes with the metal ions in solution (Pagnanelli, 2003).

In recent decades there has been a greater interest and concern about environmental issues. This has motivated the development of materials for low cost, wide availability and good adsorption capacity. At this particular point modification of adsorbent has been used in some cases with great success, with the purpose of increasing the adsorption capacity of these materials. Cellulose, hemicellulose and lignin are compounds that have structures with a large amount of hydroxyl groups. The availability of these groups are associated with good capacity to adsorb heavy metals such as Cu (II), Zn (II), Cd (II), Pb (II) among others. An interesting aspect is that these materials can be easily modified by introducing new functional groups.The modification reactions often employ a bioadsorbente polymerization reactions (Anirudhan & Noeline, 2005), functionalization with carboxylic groups, amines, amides (Shiba & Anirudhan, 2005) among others.

1.5 Electroplating wastewater

The industry of metal finishing and electroplating units are one of the major sources of pollutants which contribute greatly to the pollution load of the receiving water bodies and therefore increase the environmental risks

The industry of metal finishing and electroplating are one of the majorsource of heavy metals (Zn, Cu, Cr, …) and cyanide (Monser & Adhoum, 2002, 2002, Low & Lee,1991.). in the world. With the development of electroplating, the quantity of the electroplating wastewater have increased so fast in the last years. The heavy metals must be removed from wastewaters before discharge, as they are considered persistent, bioaccumulative and toxic (Sankararamakrishnan et al., 2007; Gupta, 2008), causing a serious threat to public health.

Heavy metal pollution around plating factories has been associated with the expansion of the plating industry in developing countries (Morgan & Lee, 1997 and Brower et al., 1997). Wastewater from the plating factories is divided into two types, one from plating manufacturing process and from rinsing process. In developed countries, removal of heavy metals in wastewater is normally achieved by advanced technologies such as ion exchange resins, vacuum evaporation, crystallization, solvent extraction and membrane technologies (Regel-Rosocka, 2010, Agrawal & Sahu, 2009, Nagarale, 2006, Ulbricht,2006). However, in developing countries, these treatments cannot be applied because of technical levels and insufficient funds. Therefore, it is desired that simple and economic removal methods to be utilized in developing countries could be established. Although chemical precipitation and coagulation–flocculation have been widely used to treat electroplating wastewater, their drawbacks like excessive chemicals consumption, sludge production, and impossibility of directly reusing heavy metals are obvious. On the other hand, adsorption methods such as ion exchange and membrane separation are simple methods for the removal of heavy metals. However, there is a limit in the generality in developing countries because chelating and ion-exchange resins are expensive. Certainly, the cost plays an important and crucial role for determining which one is to be applied. Therefore, it is worthwhile to develop economic adsorbents of heavy metals which can be generally utilized in developing countries. Consequently, in the last decades alternative adsorbents for the treatment of heavy metal contamination have been investigated.

Cyanide is capable of forming a complex with almost any metal and resulting metal complexes. They are classified according to the strength of the metal–cyanide bond through the pH at which dissociation happens. Depending on the type of metal, some simple cyanides can dissolve in water forming metal ions and cyanide ions (equation 1):

$$M_y(CN)_x \leftrightarrows M_y^- + x(CN)^-\tag{1}$$

The solubility is influenced by pH and temperature (Botz et al., 1995) and the presence of other ligants as ammonia, for example (Franson, 1992).

Cyanide complexes can generally be described by the formulae AyM(CNx), where A is an alkali or alkali-earth, y is the number of ions of A present, M is normally a transition metal, and x is the number of cyano-groups (Klenk et al., 1996).

Solution of metal ions, which already contain metal cyanide complexes, can replace the metal in the complex. The replacement depends on the respective formation constants. The metal ion in solution may also form a multimetal cyanide complex which may then precipitate as the complex, usually as an insoluble hydroxide or carbonate (Botz et al., 1995).

In the aqueous phase the equilibrium speciation of Cu(I) in cyanide solution can be represented by reactions (1)-(6) where K_1–K_4 are the equilibrium constants (here they are selected as 3.16×10^{19}, 3.39×10^4, 4.17×10^4 and 50.1, respectively (Lu et al., 2002).

$$Cu^+ + CN^- \leftrightarrows CuCN \quad K_1 = \frac{[CuCN]}{[Cu^+][CN^-]} = 3.16x10^{19}\tag{2}$$

$$CuCN + CN^- \leftrightarrows CuCN_2^- \quad K_2 = \frac{[CuCN_2^-]}{[CuCN][CN^-]} = 3.39x10^4\tag{3}$$

$$CuCN_2^- + CN^- \leftrightarrows CuCN_3^{2-} \quad K_3 = \frac{[CuCN_3^{2-}]}{[CuCN_2^-][CN^-]} = 4.17x10^4\tag{4}$$

$$CuCN_3^{2-} + CN^- \leftrightarrows CuCN_4^{3-} \quad K_4 = \frac{[CuCN_4^{3-}]}{[CuCN_3^{2-}][CN^-]} = 50.1\tag{5}$$

$$CuCN \leftrightarrows Cu^+ + CN^- \quad K_{sp(5)} = [Cu^+][CN^-] = 10^{-20}\tag{6}$$

$$Cu_2O + H_2O \rightarrow 2Cu^+ + 2OH^- \quad K_{sp(6)} = [Cu^+]^2[OH^-]^2 = 10^{-29.5}\tag{7}$$

In the aqueous phase the equilibrium speciation of Zn(II) in cyanide solution can be represented by reactions (7)-(10) with the stability constant.

$$Zn^{+2} + CN^- \leftrightarrows [Zn(CN)^+] \quad \beta_1 = \frac{[CuCN]}{[Cu^+][CN^-]} = 10^{5.34}\tag{8}$$

$$Zn^{+2} + 2CN^- \leftrightarrows [Zn(CN)_2] \quad \beta_2 = \frac{[Zn(CN)_2]}{[Zn^+][CN^-]^2} = 10^{11.97}\tag{9}$$

$$Zn^{+2} + 3CN^- \leftrightarrows [Zn(CN)_3^-] \quad \beta_3 = \frac{[ZnCN_3^-]}{[Zn^+][CN^-]^3} = 10^{16.05} \tag{10}$$

$$Zn^{+2} + 4CN^- \leftrightarrows [Zn(CN)_4^{2-}] \quad \beta_3 = \frac{[Zn(CN)_4^{2-}]}{[Zn^+][CN^-]^4} = 10^{19.62} \tag{11}$$

The diagram of the Fig. 1 shows the speciation for the copper-cyanide-water system through the pH at which dissociation happens.

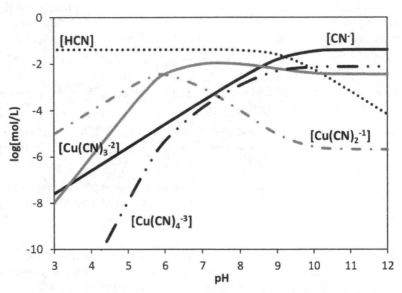

Fig. 1. Speciation diagram for the copper-cyanide-water system to 25°C ;[Cu]=1.15x10⁻²; [CN⁻]= 4.2x10⁻² (K₁=3.16x10¹⁹; K₂= 3.39x10⁴; K₃ = 4.17x10⁴; K₄= 50,1)

1.6 Metal cyanide complex on electroplating waterwaste

Cyanide, a poisonous substance dangerous to humans, animals, plants and aquatic organisms, can be found in the effluent of several industries such as paint and ink formulation, petroleum refining, explosives, case hardening, automobile industry, chemicals industry, pesticides industries, synthetic fiber production, electroplating, thermoelectric power stations, mining, electronics, and coke (Akcil, 2003; Clesceri et al., 1998; Dash et al., 2008; Monteagudo, 2004; Abou-Elela, 2008; Saarela et al., 2005; Dutra et al., 2008; Park et al., 2008; Lanza & Bertazzoli, 2002; Han et al., 2005; Mudder & Botz, 2004; Yazıcı, 2007; Zvinowanda et al., 2008; Fernando et al., 2008, Monser & Adhoum, 2002). Its effects on the human and the environment have been well reviewed by Mudder and Botz (2004). Due to its toxic nature, cyanide must be recovered, removed or destroyed.

The use of metals cyanide baths in the electroplating industry generates a strong concern related to environmental impacts due to high cadmium and cyanide toxicity (Smith and Mudder, 1991; Gijzen,2000; Waalkes, 2000). To minimize these environmental impacts wastewater treatment process are employed. The treatment intends to eliminate these toxic substances before their discharge in the environment (Butter et al., 1998). Several different cyanide removal technology are shown on table 2

Method	Description	Advantage	Disadvantage	Reference
Alkaline chlorination	Cyanide is oxidized to cyanate using chlorine or hypochlorite	Low cost when	These methods do not destroy the pollutants completely. Complete destruction of cyanate is difficult; It is less effective for iron cyanides; cannot recover cyanide; remaining chloramines and free chlorine lead to secondary contamination	(Freeman,1989; Han et al, 2005)
Hydrogen peroxide oxidation	Cyanide is oxidized to cyanate using hydrogen peroxide in the presence of copper ion	High Efficiency; Reactions involved in the process are very fast	Hydrogen peroxide is hazardous and expensive. Requires specialized equipment increasing the total capital cost. The treatment process generates ammonia, which is toxic to fish.	(Han et al, 2005)
Activated carbon adsorption	Cyanide is removed by activated carbon adsorption	Activated carbon performs both as an adsorbent and as a catalyst for the oxidation of cyanide. The adsorptive properties of activated carbons result from their high surface area and high degree of surface reactivity.	Regeneration of activated carbon is difficult	(Behnamfard & Salarirad,2009; Han et al, 2005)
Photochemical destruction	Cyanide is destroyed using ultraviolet (UV) irradiation	It ensures destruction of both free and some cyanide complexes, with weak-acid dissociable metals such as copper, without creating toxic residue	Usually inadequate by itself and requires chemical treatment, The wastewater must have a low concentration of organic matter	(Barakat et al, 2004;Han et al, 2005)
Ion exchange	Cyanide is removed by anion exchange resin	High selectivity of resins to metal cyanide complexes; Absence of hydrocyanic acid vapor which is hazardous to health.	Regeneration of resin is difficult since there are cyanide complexes besides free cyanides	(Kurama &,Catalsarik, 2000; Han et al, 2005)
Gas-filled membranes	HCN transfer through a gas-filled hydrophobic microporous membranes to striping solution containing NaOH	Cyanide can be recovered. No secondary pollutants produced. Energy and chemical requirements are low. Simple operation	The membranes define the process costs	(Han et al, 2005)

Chemical precipitation	It's recommender ferrous sulfate addition to the cyanide-containing wastewater as a simple and efficient treatment process	Economic for small-scale operation; Easy to handle by unskillful labor.	Generates solid waste	(Ismail et al,2009)

Table 2. Cyanide removal technology (Adapted from Han et al, 2005)

1.7 Technologies for the removal of cyanides

As cyanides are produced regularly by industries in large quantity in waste water streams, it is a potent health hazard for human and ecosystem. Cyanides can be removed from industrial wastes by biodegradation, physical and chemical methods (Patil and Paknikar, 2000; Ebbs, 2004). There are many reported processes for treating cyanide containing effluents (Monser & Adhoum, 2002; Dutra et al., 2008; Park et al., 2008; Mudder & Botz, 2004; Lanza & Bertazzoli, 2002; Han et al., 2005; Fernando et al., 2008; Yazıcı, 2007; Zvinowanda et al., 2008). These processes can involve biodegradation; adsorption on activated carbon; oxidation via chemical, electrochemical or photochemical processes (Monteagudo *et al.*, 2004; Saarela *et al.*, 2005; Dutra et al., 2008; Lanza & Bertazzoli, 2002; Yazıcı, 2007; Monser & Adhoum,2002); chemical precipitation (Park et al., 2008); hollow fiber gas membranes (Han et al, 2005); ultrasonic waves (Yazıcı, 2007); ion exchange (Fernando et al, 2008). The suitability of any of the above-mentioned processes to a specific cyanide-containing effluent depends on the effluent flow rate, cyanide concentration, associate chemical species, permissible level of cyanide in the effluent after treatment, technical level of the entity's employees and the economy and finances of the process.

2. Structural characterization of the biomass

There are several methods of elucidate the microstructure of the adsorbent materials from biomass. In this text, it will be present the most used by previous works in lignocellulosic biomass: Infrared spectroscopy (IR), X-Ray Diffraction (XRD) and Scanning Electron Microscopy (SEM). The macrocomponents which form the lignocellulosic fibers are cellulose, hemicellulose, lignin, pectin, wax and soluble substances, being the first three components responsible for the physical and mechanical proprieties of these fibers (Georgopoulos et al., 2005; Morán et al., 2008). This group includes wood agricultural crops, like jute, agricultural residues, such as sugar cane bagasse or corn stalks, banana fibers and other plant substances. Any lignocellulosic can be chemically modified (mercerization, acid treatment etc.) to enhance adsorption efficiency properties. This provides incentive for producing a variety of value-added products from different raw materials combined to provide improvements in cost or performance, or both (Gilbert, 1994).

2.1 Infrared spectroscopy

IR spectroscopy is a technique based on the vibrations of the atoms of a molecule. An infrared spectrum is commonly obtained by passing infrared radiation through a sample and determining what fraction of the incident radiation is absorbed at a particular energy. The goal of the basic infrared experiment is to determine changes in the intensity of a beam of infrared

pH	Technology	Pollutant	CN⁻ (mg.L⁻¹)	Efficiency (%)	Reference
10	Oxidation/Application of ferrate(VI)	Cyanide-copper-nickel system	26	98.96	(Seung-Mok & Tiwari, 2009)
10	Adsorption process/pistachio hull	Cyanide	100	99.00	(Moussavi &. Khosravi et al, 2010)
-	Adsorption process/Modified activated with carbon tetrabutyl ammonium (TBA)	Cyanide-copper-zinc system	40	73.00	(Monser &. Adhoum et al, 2002)
-	Adsorption process/Modified activated with carbon tetrabutyl ammonium (TBA)	Cyanide	40	52.5	(Monser &. Adhoum et al, 2002)
11	POA/TiO₂ and 200W(of UV irradiation)	Cyanide	26	99.40	(Barak et al, 2004)
10	Electrochemical processes	Cyanide-copper system	247	83.40	(Szpyrkowicz et al, 2000)
5.3	Integrated coagulation–gas-filled membrane absorption	Cyanide	1000-3500	> 98	(Shen et al, 2006)
9-11	Solvent extraction	Cyanide-copper system	1100	99	(Alonso-González et al, 2010)
3.2	Adsorption process	Cyanide	-	98	(Moussavi ,2011)
8.0	Chemical precipitation	Cyanide-zinc system	18	99.47	(Ismail et al,2009)

Table 3. Technologies for the removal of cyanides

radiation as a function of wavelength or frequency. The energy at which any peak in an absorption spectrum appears corresponds to the frequency of a vibration of a part of a sample molecule (Stuart, 2004). Thus, an infrared absorption spectrum of a material is obtained simply by allowing infrared radiation to pass through the sample and determining what fraction is absorbed at each frequency within some particular range (Bower & Maddams, 2006).

Some examples of IR spectra of the lignocellulosic material are shown in Fig. 1. These samples had been studied in previous works for removal of toxic metal ions from aqueous industrial effluents or as composite materials: Sugar cane bagasse (Sousa et al., 2009) and fibers: coir (Esmeraldo et al., 2010), sisal (Barreto et al., 2009), jute (Barreto et al., 2010a) and Banana (Barreto et al., 2010b). It can be observed that the components of biomass are most likely consisted of alkenes, esters, aromatics, ketenes and alcohol, with different oxygen-containing functional groups (Yang et al., 2007). The main lignocellulosic IR vibrational modes presented are from cellulose, hemicelluloses and lignin. All samples presented two main transmittance regions. The first one at low wavelengths in the range 1800-500cm⁻¹ and the second one at higher wavelengths corresponding to the 3700-2750cm⁻¹, approximately. However, there are modification in signal intensity of these spectra due to different cellulose, hemicelluloses and lignin concentrations of each material. A resume of the assignment of main IR bands in these materials was building with dates from literature (Esmeraldo et al., 2010; Barreto et. al., 2011, Barreto et al., 2010a; Barreto et al., 2010b,Yang et al., 2007; Viera et al., 2007; Morán et al., 2008; Bilba et al., 2007) and it is presents in Table 2.

The spectrum of the Fig. 2 exhibited O–H stretching absorption of around 3430 cm⁻¹, C–H stretching absorption of around 2920 cm⁻¹, C=C benzene stretching ring of around 1634 cm⁻¹

and C–O–C stretching absorption of around 1058 cm⁻¹. There are other bands with weak and very weak signal that are described in Table 4. These absorptions are characteristic of the common lignocellulosic fiber.

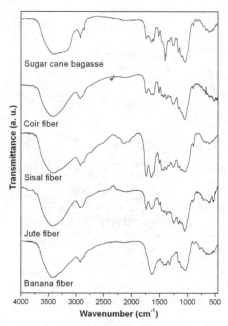

Fig. 2. Infrared spectra of natural lignocellulosic material.

Wavenumber (cm⁻¹)	Functional group
3600-3100	OH stretching
2970-2860	CH stretching of CH_2 and CH_3 groups
1765-1700	C=O stretching of acetyl or carboxylic acid
1634	Carbonyl stretching with aromatic ring
1620–1595	C=C stretching of aromatic ring (lignin)
1512	C=C stretching of aromatic ring (lignin)
1429	CH_2 bending
1376	C-H deformation
1335	OH in plane bending
1250	C-O stretching of ether linkage
1166	C-O-C antisymmetric bridge stretching
1062	C-O symmetric stretching of primary alcohol
904	β-Glucosidic linkages between the sugar units
900-700	C-H Aromatic hydrogen (lignin)
700-400	C-C stretching (lignin)

Table 4. Assignment of main functional groups of lignocellusosic materials.

The infrared spectroscopy was used to analyze the effect of the chemical treatment on the surface structure of the fiber. For example, Barreto and co-workers (Barreto et. al., 2011) observed that in sisal fibers, the weight loss due to the partial dissolution of hemicellulose, lignin and pectin are clearly identified in the band in 1730 cm-1, which disappears when the fibers are treated by a NaOH aqueous solution. This vibration mode was only observed to raw sisal fibers. Another important consideration is the broadness of the hydroxyl band at 3500–3300 cm-1. For the natural sisal fibers the band is at 3419 cm-1, characteristic of the axial vibration of hydroxyls from cellulose (carbons 2, 3 and 6 of the glucose) (Esmeraldo, 2006; Calado et. al., 2000) and it is showing a broadness as a function of the chemical treatment and due to changes of the inter- and intra-molecular hydrogen bonding in polysaccharides (Esmeraldo et al., 2010), which reaches a frequency of 3372 cm-1 for the fiber treated with NaOH 10%. However, for sugar cane bagasse study presents by Sousa and co-workers (Sousa et al., 2009), it was observed that acid treatment (1.0M HCl) removed or decreased some modes of lignin: absorptions due to C–Hn (alkyl and aromatic) stretching vibrations (2918 and 2850 cm-1), absorptions characteristic of the C=O stretching vibration (1708 cm-1), and a signal typical of an aromatic skeleton (1604 cm-1). It can also be noted that the intensities of almost all bands were lower after acid treatment. Thus, IR spectroscopy is an important tool to evaluate the changes modification of the biomass structure before and after to the chemical treatment.

2.2 X-ray diffraction

XRD is the most widely used technique for general crystalline material characterization. It is noncontact and nondestructive, which makes it ideal for in situ studies (Brundle et al., 1992). It is routinely possible to identify phases in polycrystalline bulk material and to determine their relative amounts from diffraction peak intensities. The XRD measurements are used to examining crystallite solid as ceramics, metals, geological materials, organics and polymers. The samples for measurements may be single crystals, powders, sheets, films and fibers.

Cellulose is the main renewable carbon source in nature. It presented a high organized state nature (fibrous crystal) due to its linear polymer arrange of pure anhydroglucose units connected by 1,4β-glucosidic bonds (Chang et al., 1981). Each residue is rotated 180° compared with its neighbors. The degree of polymerization (DP) of native cellulose is in the range of between 7,000 and 15,000, where DP = Molecular weight of cellulose / Molecular weight of one glucose unit. It occurs in crystal and noncrystal regions as well as in association with lignin deposition in the secondary wall. The cellulose chains are oriented in parallel and form highly organized crystalline domains interspersed by more disorganized, amorphous regions. Cellulose chains form numerous intra- and intermolecular hydrogen bonds, which account for the formation of rigid, insoluble microfibrils (Buckeridge et al., 2011). The native crystalline form has a structure designated as type I, which can be converted into type II by alkaline treatment. Fig. 3 shows a representative model of the

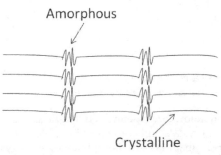

Fig. 3. Fibrillar model of cellulose.

molecular orientation in the crystallite. One can observed that there are intermittent highly ordered areas labeled crystalline regions, separates by less ordered or amorphous regions. This characteristic is important to understand the degradation mechanism of lignocellulosic material. For instance, amorphous regions of the cellulose are first hydrolyzed followed by hydrolysis of crystalline regions at a much slower rate.

Fig. 4 shows the XRD patterns of the untreated and acid treated sugar cane bagasse. Three peaks are presented in the directions of 15.2^0, 16.5^0 and 22.4^0. They assign the cellulose standard profile from JCPDS (Joint Committee on Powder Diffraction Standards, 1986). Ouajai and Shanks (Ouajai & Shanks, 2005) observed these same directions in hemp fiber and labeled them as 101, 10$\bar{1}$ and 002 diffraction planes from cellulose crystalline phase. Besides this phase, there was an amorphous phase characterized mainly as lignin. This polymer is associated to the cellular wall, conferring mechanical strength to the fiber, and when its concentration increases, the crystalline fraction decreases. The sugar cane bagasse fibers contain approximately 46% cellulose, 25% hemicelluloses and 21% lignin as the main components (Buckeridge et al., 2011). It also observed three other peaks on 26.52^0, 29.36^0 and 30.84^0 for both samples that could be attributed to impurities such as SiO_2 and $CaCO_3$, whose intensities decreased after acid treatment.

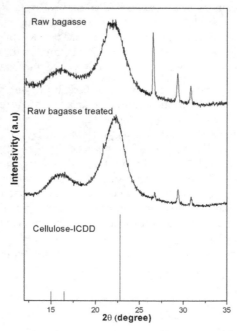

Fig. 4. XRD of the raw sugar cane bagasses and treated with 1.0M HCl, showing the crystalline and amorphous phases obtained by in comparison to data from ICDD.

The diffraction pattern from amorphous materials (including many polymers) is devoid of the sharp peaks characteristic of crystals and consists of broad features or halos. Many polymers, as biomass used to adsorbent material, are amorphous or semicrystalline, and for polymeric materials, XRD is used to clarify the structure, morphology, and degree of crystallinity. The sugar cane bagasse crystalline fraction (F) can be obtained through separation and integration of crystalline and amorphous peaks areas under the diffraction X-rays plot, by the Equation (12):

$$F = \frac{\sum I_C}{\sum (I_C + I_A)} x100 \tag{12}$$

where, I_C is the difractogram total crystalline phase area, and the I_A is the difractogram total amorphous phase area. For XRD presented in Fig. 3, the F value for the untreated material (62%) was close to the acid-treated material (67%) (Sousa et al., 2009). However, depend on biomass, the F value can present results in different ways. As it was observed for the banana fiber (Barreto et al., 2010b): the raw material presented 63.5% and after mercerization treatment (NaOH 0.5%), this value increased to 79.2%. This difference was due to the lignin be partially removed in alkaline solution, i. e., the banana fiber lost part of its amorphous component. Thus, the F values depend on the natural concentrations of lignin, cellulose and hemicelluloses presented by biomass and chemical treatment efficiency to remove amorphous phases.

2.3 Scanning electron microscopy

The SEM provides the investigator with a highly magnified image of the surface of a material that is very similar to what one would expect if one could actually "see" the surface visually (Brundle et al., 1992). SEM images the sample surface by scanning it with electron beams in a raster scan pattern. The electrons interact with the sample atoms producing signals that contain information about the sample's surface topography, composition and other properties such as electrical conductivity (Liu et al., 2010).

Not only is topographical information produced in the SEM, but information concerning the composition near surface regions of the material is provided as well. Additionally, the SEM can also be used to provide crystallographic information. Surfaces that to exhibit grain structure (fracture surfaces, etched, or decorated surfaces) can obviously be characterized as to grain size and shape. Thus, SEM can be used to materials characterization: topographical imaging, Energy-Dispersive X-Ray analysis and the use of backscattering measurements to determine the composition of the systems.

The surface morphology and porosity of natural fiber have been recognized as significant factors for composite interfaces, and their effects on the performance of composites have been investigated (Han & Choi, 2010). This information about is also important for an adsorbent. It is possible observe the grain or fiber size, porosity and morphology and compare to before and after some chemical treatment employed. The increase or decrease to adsorption capacity can be the answer in this way. Thus, surface morphology and microstructure of the adsorbents can be studied by SEM. For biomass, it is necessary coat the sample with a conductor material, as gold, platinum or with a layer of carbon. This happens due to slow electron conductivity of the kind of the material and this affects the images quality.

Fig. 5(a) shows the SEM morphology at raw banana fiber. It is observed a regular structure with discrete net fibrils due to the presence of hemicelluloses, lignin, and wax, where these constituents confer mechanical strength to the natural composite. After alkali treatment

(NaOH 1%), there is considerable structure modification as a visible separation of the fibrils, as are shown in Fig. 5(b). This happened due to remove partially of the some components. One of the aims of this treatment was to increase the surface area and decrease the hydrophilic groups. This hydrophilic nature can lead to incompatibility and poor wettability in a hydrophobic polymer matrix, and weak bonding in the fiber/matrix interface. Some cavities also appeared due to modifications on the surface.

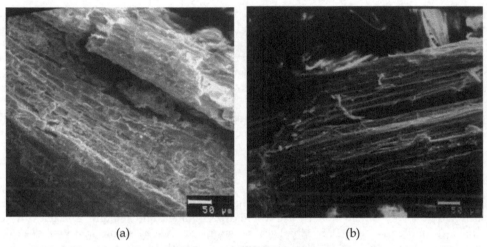

(a)   (b)

Fig. 5. SEM of the (a) raw banana fiber and (b) after alkali treatment with NaOH 1%

In other example, it is presented the raw sugar cane bagasse microstructure, see Fig. 6 (a). This material presents a reasonable homogeneity in aggregate shape. However, after acid treatment with HCl, some impurities were removed Fig. 6(b). It was also evidenced exposure and dispersion of the fibers, resulting in increased roughness. This procedure increases the superficial area of the sugar cane bagasse fiber and contributes to improve the metal adsorption capacity. Consequently, this fiber can be used as an efficient natural material to remove toxic metal ions from electroplating industry wastewater. A comparative study between metal adsorption behavior and lignocellulosic biomass morphology was

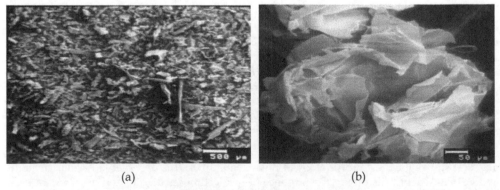

(a)   (b)

Fig. 6. SEM of the raw sugar cane bagasse (a) and after acid treatment with HCl (b).

presented in previous studies (Bower & Maddams, 2006; Barreto et al., 2010b). The important observation is that biomass chemical modified was more efficient than raw material in the metal removal from aqueous solution. Additionally, this kind of material, when used to adsorption material, has the advantage of low density, low cost, recyclability and biodegradability.

3. Adsorption of single and multi-metal ions

3.1 Adsorption kinetics

The study of the time dependence of adsorption on solid surface (adsorption kinetic) help in predicting the progress of adsorption in industrial application, however the determination of the adsorption mechanism is also important for design purposes. In a solid-liquid adsorption process, the transfer of the adsorbate is controlled by either boundary layer diffusion (external mass transfer) or intra-particle diffusion (mass transfer through the pores), or by both (Dabrowski, 2001). It is generally accepted that the adsorption dynamics consists of three consecutive steps:

- Transport of adsorbate from the bulk solution to the external surface of the adsorbent by diffusion through the liquid boundary layer.
- Diffusion of the adsorbate from the external surface and into the pores of the adsorbent.
- Adsorption of the adsorbate on the active sites on the internal surface of the pores.

The last step, adsorption, is usually very rapid in comparison to the first two steps. Therefore, the overall rate of adsorption is controlled by either film or intra-particle diffusion, or a combination of both. Many studies have shown that the boundary layer diffusion is the rate controlling step in systems characterized by dilute concentrations of adsorbate, poor mixing, and small particle size of adsorbent (Singh & Mohan, 2004).

In order to investigate the adsorption kinetics of metal ions adsorption on adsorbent, pseudo-first order, pseudo-second order, and intraparticle diffusion models are the models more frequently applied, which are pseudo-first-order (Lagergren, 1898), pseudo-second-order (Ho & Mckay, 1999) and intraparticle diffusion model (Weber & Morris, 1963), expressed as:

$$\log(q_e - q_t) = \log q_e + \frac{K_1}{2.303} \times t \tag{13}$$

$$\frac{t}{q_t} = \frac{1}{K_2 \, q_e^{\,2}} + \frac{1}{q_e} \times t \tag{14}$$

$$q_t = k_{id} t^{1/2} + C \tag{15}$$

Where, q_e and q_t are the adsorption capacities (mg g^{-1}) at the equilibrium and at a time determined, respectively, t the time, while K_1 (min^{-1}) and K_2 (mg g^{-1} min^{-1}) are rate constants related to the first order and second order models. C is the intercept and k_{id} is the intra-particle diffusion rate constant (mg g^{-1} min$^{-1/2}$), which can be evaluated from the slope of the linear plot of q_t versus t $^{1/2}$. The values C provide information about the thickness of the boundary layer, the larger the intercept, the greater the boundary layer effect (Kavitha & Namasivayam, 2007).

If intra-particle diffusion occurs, then q_t versus $t^{1/2}$ will be linear and if the plot passes through the origin, then the rate limiting process is only due to the intra-particle diffusion. Otherwise, some other mechanism along with intra-particle diffusion is also involved.

The typical results obtained by Sousa et al., 2010; Moreira et al., 2010 for the metal ions sorption onto lignocellusic adsorbents by investigating the pseudo first and pseudo-second-order rate equations order and intraparticle diffusion model are presented in Table 5. Applications of these rate equations for describing the kinetics studies were investigated for batch technique due its simplicity. For this, erlenmeyer flasks of 50mL containing 0.30 g of adsorbent with 10 mL of multi-metal solution (100 mg L^{-1}, at pH 5.0) were shaken (at 300 rpm) in room temperature, at 28°C. After a predetermined time, volumes of the solution were removed, filtrated and analyzed. The amount of adsorption was calculated at equilibrium achieved in 4-5 h by equation 16.

$$q_e = \frac{(C_o - C_e)V}{m}$$

(16)

Where: C_o is the solution initial concentration (mg L^{-1}); C_e is the equilibrium concentration of adsorbate (mg L^{-1}); V is the solution volume (L) and m is the mass do adsorbent (g).

The linear plot of log (q_e-q_t) versus t for pseudo first order, $1/q$ versus t for pseudo-second order and q versus $t^{1/2}$ for intraparticle diffusion models are used for constants K_1, K_2 and K_f calculated from slope of the corresponding linear equations and correlation coefficients (R^2). From results presented in Table 5 can be noted that the experimental q_e values show a reasonable agreement with the calculated q_{cal} value by first order equation, indicating that the adsorption system belongs to the first order reaction. In contrast, do not agree between experimental and calculated q_{cal} values by second-order equation is observed. This indicates that the adsorption of multi-metal on the sugar cane bagasse is not described by a second order reaction.

When the intraparticle diffusion model can be applied then the plots of q versus $t^{1/2}$ is linear, and if it passes thought the origin the limiting processes is only due to intraparticle diffusion (Weber & Morris, 1963). However, since that the plots of q versus $t^{1/2}$ (not showed) was linear but is not pass through origin, then the intraparticle diffusion is not the only rate limiting mechanism on multi-metal adsorption on the cane sugar bagasse studied.

Table 5 presented the results obtained of adsorption kinetics studies for Cu^{2+}- Ni^{2+} -Cd^{2+} on the sugar cane bagasse (Sousa et al.,2010). It is observed that the experimental q_e values are reasonable in agreement with the calculated q_c value for first order equation, indicating that the adsorption system belongs to the first order reaction. In contrast, do not agree between experimental and calculated q_e values by second-order equation can be observed. This suggest that adsorption of Cu^{2+}- Ni^{2+} - Cd^{2+} on the sugar cane bagasse not follows a second order reaction. However, the results found by Moreira et al., 2010 showed that Cu^{2+}, Ni^{2+} and Cd^{2+} adsorption on cashew bagasse has a good agreement between the experimental (q_e) and the calculated (q_{ca}) values, as shown in Table 5. The high values for coefficient of correlation, R^2, indicate that there is strong evidence that the Cu^{2+}, Ni^{2+} and Cd^{2+} adsorption onto cashew bagasse follows the pseudo- first and second order kinetic models.

When the intraparticle diffusion model can be applied then the plots of q versus $t^{1/2}$ is linear, and if it passes thought the origin the limiting processes is only due to intraparticle diffusion

(Weber & Moris, 1963). However, since that the plots of q versus $t^{1/2}$ (not showed) was linear but is not pass through origin, then the intraparticle diffusion is not the only rate limiting mechanism on multi-metal adsorption on the cane sugar and cashew bagasse studied. The values correlation coefficient (R^2) obtained for the plots were not satisfactory.

Metal	First order					Second order			Intraparticle Diffusion	
	C_o (mg L^{-1})	$q_{e(exp)}$ (mg g-1)	$q_{(cal)}$ (mg g^{-1})	K_1 (min^{-1})	R^2	$q_{(cal)}$ (mg g^{-1})	K_2 (mg g^{-1} min^{-1})	R^2	K_f (mg g^{-1} min$^{-1/2}$)	R^2
			sugar cane bagasse							
Cu^{2+}	74.14	2.015	2.01	0.98	0.890	3.11	0.140	0.88	0.79	0.860
Ni^{2+}	108.81	0.827	1.05	0.61	0.890	-	-	-	0.29	0.810
Cd^{2+}	117.99	2.54	2.56	0.51	0.920	5.44	0.040	0.93	0.51	0.860
			Cashew bagasse							
Cu^{2+}	102.18	1.982	1.933	0.016	0.956	1.95	0.518	1.00	0.009	0.779
Ni^{2+}	82.93	1.588	1.573	0.006	0.790	1.59	0.873	1.00	0.003	0.755
Cd^{2+}	90.70	1.803	1.800	0.012	0.852	1.80	0.022	1.00	0.001	0.909

Table 5. Parameter of adsorption kinetic in the sugar cane bagasse and cashew bagasse in multi-metal solutions (10-200 mg.L^{-1}) at pH 5.0 , time equilibrium achieved in 4-5 h.

3.2 Adsorption isotherm equilibrium

Most experimental and theoretical studies of the adsorption at solid-liquid interface have been carried for single or multi metal ions removal from aqueous solution by diverse types of low cost adsorbents (Pasavant et al., 2006; Aksu et al.,2002; Cay et al., 2004; Amarasinghe & Williams, 2007). In the practice, the adsorption metal ions from wastewater generally involve the simultaneous presence of metal ions in wastewater promoting a competition between different metal ions by adsorption site. However, study of equilibrium modeling of multi-metal ions is essential for understand the real system, but usually are neglected (Febrianto, 2009).

Adsorption is an important process that describes the interaction between adsorbent and metal ion to develop design model for wastewater industrial treatment. The applicability of relationship between the experimental adsorption capacities and the metal ions concentrations (adsorption isotherm) have been widely used by the Langmuir and Freundlich models (Frebianto et al., 2009).

3.2.1 Langmuir model

The Langmuir adsorption model is based on the assumption that maximum adsorption corresponds to a saturated monolayer of solute molecules on the adsorbent surface, with no lateral interaction between the adsorbed metal. The Langmuir adsorption isotherm model is successfully used to explain the metal ions adsorption from aqueous solutions (Langmuir, 1918). The expression of the Langmuir model is given by Eq.(17),

$$q_e = \frac{q_{max}.K_L.C_e}{(1 + K_L C_e)} \tag{17}$$

and the linearized Lagmuir isotherm equation can be expressed as:

$$\frac{1}{q_e} = \frac{1}{q_{max}} + \left(\frac{1}{q_{max}K_L}\right)\left(\frac{1}{C_e}\right) \tag{18}$$

3.2.2 Freundlich model

The Freundlich isotherm gives the relationship between equilibrium liquid and solid phase capacity based on the multilayer adsorption (heterogeneous surface). This isotherm is derived from the assumption that the adsorption sites are distributed exponentially with respect to the heat of adsorption and is given by Freundlich, 1906).

The Freundlich isotherm is an empirical equation employed to describe heterogeneous systems. The Freundlich equation is expressed as:

$$q_e = K_F C_e^{1/n} \tag{19}$$

and linearized Freundlich isotherm equation can be expressed as:

$$\log q_e = \log K_F + 1/n \log C_e \tag{20}$$

Where q_e (mg.g^{-1}) is the amount of metal ion adsorbed, expressed as mg metal ions per g adsorbent, C_e (mg.mL^{-1}) the equilibrium concentration of metal ion in solution, q_{max} (mg.g^{-1}) and K_L (L.mg^{-1}) are constants of Langmuir related to the maximum adsorption capacity (mg.g^{-1}) and heat of adsorption, respectively, while K_F (mg$^{1-(1/n)}$.(g^{-1}).L$^{1/n}$) and $1/n$ are constants of Freundlich related to the adsorption capacity and to surface heterogeneity, respectively.

3.2.3 Applications of Langmuir and Freundlich models

The linear fit of the experimental data using the Langmuir and Freundlich models permitted to obtain correlation coefficients greater than 0.90. Although its satisfactory value indicate, the applicability of both models is very difficult to identify the adsorption equilibrium model which represented the experimental data most correctly considering only the correlation coefficients. Hence, a parameter known as normalized percent deviation, P, (Ayranci, 2005) can be applied, according to the following equation 21:

$$P = (100/N) \sum (|q_e - q_{cal}|/q_e) \tag{21}$$

where q_e is the experimental adsorption capacity, q_{cal} the predicted adsorption capacity, and N the number of observations. The lower the P value is, better the fit is.

According to the literature (Cooney, 1993), the slope of the initial curvature of an adsorption isotherm indicates whether or not an adsorption system is efficient, which, for a Langmuir isotherm, can be expressed in terms of the separation factor R_L (Cooney, 1993; Ngah et al, 2008):

$$R_L = 1 / (1 + K_L C_o) \tag{22}$$

where K_L (L mg⁻¹) is the Langmuir constant and C_o the metal ion initial concentration.

The type of isotherm is considered to be unfavorable, i.e., the solute has a preference for adsorption to the solid phase over dissolution in the liquid phase ($R_L > 1$), linear ($R_L=1$), favorable ($0< R_L < 1$) or irreversible ($R_L=0$) depending on the value of R_L (Cooney, 1993, Ngah et al, 2008). The experimental R_L values were between 0 and 1 for all initial concentrations as known in Tables 6 and 7, hence, adsorption of metals ions studied on the adsorbent is considered to be favorable.

Applications of Langmuir and Freundlich models for describing adsorption isotherms with green coconut shell powder and cane sugar bagasse (treated with sodium hydroxide) for toxic metals removal from aqueous effluents were studied (Sousa et al., 2011; Sousa et al., 2010). Equilibrium adsorption isotherms of multi-metal (Cd^{+2}-Cu^{+2}- Ni^{+2}- Pb^{+2}- Zn^{+2}) on the green coconut shells and cane sugar bagasse are given in Fig. 7. The parameters of Langmuir (K_L; q_{max}) and Freundlich (K_F; $1/n$), determined from the slope and intercept of the plots of $1/q_e$ versus $1/C_e$ and log q_e versus log C_e are shown in Tables 6 and 7.

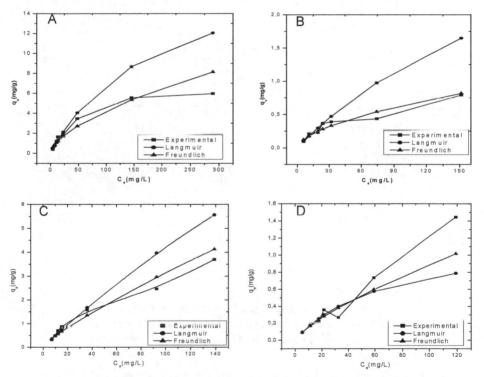

Fig. 7. Adsorption isotherm to metal ions: Pb^{+2}(a), Cd^{+2}(b), Cu^{+2}(c) and Ni^{+2}(d) on sugar cane bagasse, contact time 24h and pH 5.0.

The P values calculated in according to the Langmuir and Freundlich equations for single and multi-systems for sugar cane bagasse are given in Tables 6. It is observed that for single metal ion system the P deviations are lowest when the experimental data were fitted to the

Langmuir equation for all single metal ions studied, while the Freundlich model proved suitable for multi-metal system

The Langmuir and Freundlich isotherms for multi-metal ions adsorption on green coconut shell powder are shown in Fig. 8. The parameters determined for single and multi-metal ions each model are given in Table 3. The experimental data show that the Langmuir-type isotherms describe well the adsorption mechanism for Pb^{+2}, Ni^{+2} and Cd^{+2}, while Zn^{+2} and Cu^{+2} are of the Freundlich-type as can be observed by the values of correlation coefficients (R^2). The adsorption capacity of the multi-metals studied followed the order: $Zn^{+2} > Pb^{+2} > Cu^{+2} > Cd^{+2} > Ni^{+2}$. This can be attributed to the specificity of active sites, to varying affinities for adsorption or to competitive effects. Sekhar et al., 2003 referred that multicomponent systems, the complex interactions of several factors such as ionic charge and ionic radii will account for the differences in the metal removal capacity of the adsorbent. As a result, ordering of the metal ions based on a single factor is very difficult.

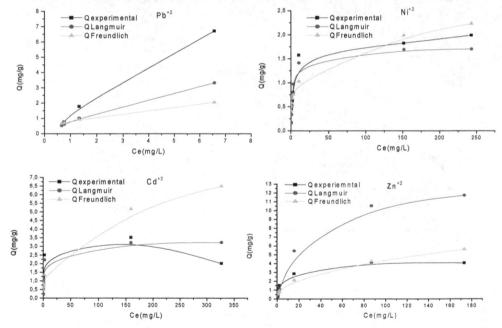

Fig. 8. Comparison of the Langmuir and Freundlich models with the experimental isotherm for Pb^{+2}, Ni^{+2}, Cd^{+2} and Zn^{+2} in a multimetal solution on green coconut shell powder treated with NaOH 0.1 mol.L^{-1}/3 h, pH 5, $C_{biomass}$ = 40 g.L^{-1}, particle size 60-99 mesh, ambient temperature (28±2 ºC).

The experimental data for single systems clearly show that the Langmuir-type isotherms describe well the adsorption mechanism Pb^{+2}, Cd^{+2} and Cu^{+2}, while Zn^{+2} is of the Freundlich-type. On the other hand, Ni^{+2} agrees well with both models. The adsorption capacity of the single metals studied followed the order: $Cd^{+2} > Zn^{+2} > Cu^{+2} > Pb^{+2} > Ni^{+2}$, which can be correlated to the ionic radius of Pauling, except for Pb^{+2} and Ni^{+2} (Table 2) (Vázques et al., 2002; Mohan & Singh, 2002; Sekhar et al., 2003; Mattuschka & Straube, 1993).

The values of R_L for single and multi-metal ions are all between 0 and 1 indicating that adsorption on green coconut shell powder is very efficient for these systems (Table 4).

Metal ions	Langmuir					Freundlich			
	K_L (L mg^{-1})	q_{max} (mg g^{-1})	R_L	P	R^2	K_F (mg$^{1-(1/n)}$ (g^{-1})L$^{1/n}$)	$1/n$	P	R^2
Single metal									
Cu^{+2}	0.006	7.34	0.400-0.952	8.52	0.98	0.072	0.82	12.73	0.96
Ni^{+2}	0.015	1.23	0.214-0.891	9.22	0.97	0.028	0.76	10.89	0.93
Pb^{+2}	0.005	19.92	0.291-0.925	10.26	0.98	0.253	0.61	17.14	0.93
Cd^{+2}	0.002	1.07	0.180-0.868	9.20	0.95	0.044	0.58	16.62	0.90
Muti-metal									
Cu^{+2}	0.06	1.10	0.062-0.689	18.97	0.89	0.080	0.64	21.54	0.77
Ni^{+2}	0.03	0.59	0.082-0.939	26.14	0.54	0.009	0.97	20.46	0.89
Pb^{+2}	0.062	12.66	0.614-0.979	23.58	0.74	0.009	1.00	18.77	0.65
Cd^{+2}	0.002	1.17	0.158-0849	29.20	0.66	0.044	0.58	16.62	0.91

R_L=separation factor

Table 6. Adsorption isotherms parameters of Langmuir and Freundlich, for single-metal adsorbed on sugar cane bagasse.

Adsorption	Langmuir				Freundlich		
	K_F (mg$^{1-(1/n)}$ (g^{-1}) L$^{1/n}$)	q_{max}(mg.g^{-1})	R_L	R^2	K_F (mg$^{1-(1/n)}$ (g^{-1}) L$^{1/n}$)	$1/n$	R^2
ingle metal							
Cu^{+2}	0.072	10.45	0.441 – 0.013	0.994	0.767	0.614	0.958
Ni^{+2}	0.110	6.71	0.339 – 0.010	0.986	0.685	0.547	0.986
Pb^{+2}	0.086	8.32	0.560 – 0.01	0.986	0.814	1.492	0.925
Cd^{+2}	0.085	17.51	0.865 – 0.154	0.979	1.315	0.669	0.880
Zn^{+2}	0.072	10.45	0.697 – 0.060	0.994	0.767	0.614	0.958
Multi- metal							
Cu^{+2}	0.227	5.09	0.354 – 0.004	0.969	0.834	0.581	0.977
Ni^{+2}	0.451	1.72	0.241 – 0.003	0.967	0.580	0.245	0.900
Pb^{+2}	0.112	7.89	0.560 – 0.008	0.972	0.812	0.497	0.886
Cd^{+2}	0.962	3.24	0.098 – 0.001	0.991	1.04	0.315	0.759
Zn^{+2}	0.045	13.3	0.677 – 0.025	0,896	0,877	0,341	0,903

Table 7. Adsorption isotherms parameters of Langmuir and Freundlich, with the correlation coefficients (R^2) for single and multi-metal adsorbed on coconut shells bagasse.

3.2.4 Prediction of multi metal equilibrium

In the practice, the adsorption metal ions from wastewater generally involve the simultaneous presence of metal ions in wastewater promoting a competition between different metal ions by adsorption site. However, study of equilibrium modeling of multi-metal ions is essential for understand the real system, but usually are neglected (Frebianto,

2009). The competitive adsorption of metal ions on the sugar cane bagasse was studied (Sousa et al., 2011) using the extended Langmuir model, expressed as:

$$q_i = \frac{q_m K_i C_i}{1 + \sum_{j=1}^{n} K_j C_j}$$ (23)

for i = 1,2,..., n, where i and j represent the metal ions, q_i and q_m they are the adsorption capacity and maximum capacity of adsorption (mg metal / g adsorbent) and K Langmuir constant.

Considering that a solid adsorbent has a given surface area, then the presence of other solutes implies a competition for available adsorption sites. In general, the presence of other solutes decreases the adsorption of any given solute (Cooney, 1993). Based on the isotherms showed in Fig. 9 (a,b) can be seen that there is a substantial difference between predicted and experimental values indicating a considerable effect of competition by metals adsorption sites. The results showed in Figure 8 prove competition for binding sites from comparison of the experimental and predicted data. Thus, adsorption is not specific and limited to a maximum binding capacity.

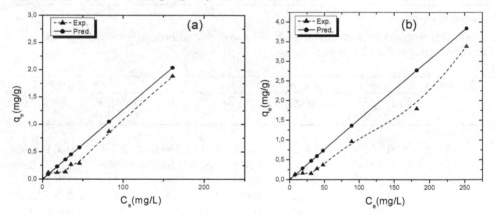

Fig. 9. Comparison between the experimental and predicted adsorption models for Ni^{+2} (a) and Cd^{+2} (b) on cane sugar bagasse treated with NaOH 0.1 mol.L$^-$/3h, in a multimetal solution. Conditions: pH 5, $C_{biomass}$ = 40g.L^{-1}, ambient temperature (28±2 ºC).

3.3 Adsorption in fixed bed – column

3.3.1 Breakthrough curves theory

The majority of adsorption studies have been carried out in the batch mode, but a fixed bed conventional system (column) should be economically most valuable for wastewater treatment (Cooney, 1999). A system of fixed bed conventional is compound of a column which particles of the bioadsorbent are putting in contact with the effluent. The column efficiency is described through of the concept of breakthrough curve. An ideal breakthrough curve is show in the Fig. 10, where C_o and V_e are the adsorbate concentration in the effluent and the volume of effluent passed in the column, respectively. The ideal curve admits that the adsorbate removal is complete above of the initial stages of operation. In this curve the breakpoint has been chosen arbitrarily at C_b and occur when the effluent concentration reached 5% of the initial concentration C_0. The column reaches the complete saturation when the concentration C_x, closely approaching C_0. The total amount of effluent, V_b, passed in the column until the breakpoint and the nature of the breakthrough curve between the values of V_b and V_x are important to design of a column (Cooney, 1999).

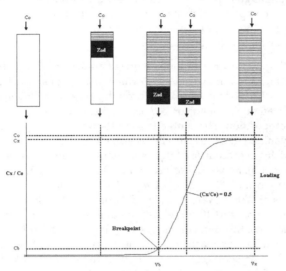

Fig. 10. Ideal breakthrough curve. (Adaptaded from Cooney, D.O.(1999). *Adsorption design for wastewater treatment* , Lewis publishers, ISBN 1566703336 9781566703338. Boca Raton, Fl and McKay, G. (1996) (Editors), Design of adsorption contatacting systems,in: Use of Adsorbents for the Rremoval of Pollutans from Wasterwaters,pp 99-131,CRC press,Inc, Boca Raton, Florida ISBN: 0-8493-6920-7

The part between C_x (exhaustion point) and C_b (breakpoint) is called the primary adsorption zone (PAZ) and the time needed for PAZ to move down the column is calculated by Equation 24 (Gupta *et al.*, 1997; 2000; Kundu & Gupta, 2005):

$$t_x = \frac{V_x}{F_m}$$

(24)

Where, t_x is the time to establish PAZ (min), F_m is the flow rate (mL/min) and V_x is the exhaustion volume (mL).

The maximum capacity of removal of toxic metal ions in the column is given by Equation 25 (Gupta *et al.*, 1997; 2000; Kundu & Gupta, 2005):

$$Q = \frac{C_0 * V}{m_s} \int_{t=0}^{t=x} (1 - \frac{C}{C_0})dt$$

(25)

Where, Q is the maximum adsorption capacity (mg/g); C_0 and C are the initial concentration of the solution and the concentration of the metal ion in a determined volume (mg/L), respectively; m_s is the mass of the adsorbent (g); V is the flow rate (L/min) and t is the time in minutes.

3.3.2 Breakthrough curves

Metal cyanides can occur of various forms in the wastewaters, generated from electroplating industries, depending of the solution pH (Patil & Paknikar, 1999; Bose et al., 2002). For wastewaters containing high copper concentration and relative concentration of zinc, nickel and cyanide, the weak complexes of nickel and zinc are present in negligible concentrations. In contrast, copper-cyanide complexes are present in appreciable quantity, due to high affinity of copper by cyanide (Bose et al., 2002).

Since the electroplating wastewater contains metal-cyanide complexes, thus the influence of these species on metal ions removal of wastewater using sugar cane bagasse have been studied (Sousa et al., 2010). For this, the breakthrough curves were investigated for synthetic effluent spiked with amount known of copper, nickel and cyanide ions similarly to the effluent industrial composition. The synthetic samples were treated by alkaline chlorination for destruction of cyanide (Akcil, 2003). The treatment of the synthetic sample was carried out with NaOCl 2M by overnight for total oxidation of cyanide (18 mg/L^{-1}) to inorganic carbon. The precipitated metal was dissolved with 1.0 mol.L^{-1} H$_2$SO$_4$ to yield a final solution with metal concentrations of 200mg.L^{-1}, 14.4 mg.L^{-1} and 65.5 mg.L^{-1} of Cu^{2+}, Ni $^{2+}$ and Zn^{2+}, respectively. Volumes of 200mL were percolated through a column (with 4.0g of adsorbent) at flow rate of 2.0 mL min^{-1} and 10 mL eluted volume was collected at the exit of the bed column to obtain the breakthrough curves given in Fig. 11 (a,b). It is seen that the column capacity at complete exhaustion reached a plateau when the sample volume was about 70.0 mL of the effluent. The percentages of saturation of the column, calculated at breakpoint were 73.7, 75.2, and 67.5% for copper, nickel and zinc respectively.

The adsorption capacities for the metal ions obtained from breakthrough curves are shown in Table 8. It is observed that copper adsorption by cane sugar bagasse is higher than zinc and nickel adsorption (in absence and presence of cyanide). The copper ion interacts strongest with cyanide to form copper-cyanide complexes (anionic) in aqueous solutions on dependency of pH. Therefore, a considerable removal of copper in solution containing cyanide, in acidic conditions, may be due to the formation from insoluble specie CuCN (Bose et al., 2002). However, the contrast can be observed for zinc and nickel, which form relatively weak complexes with cyanide in aqueous solutions at low pH. Zinc and nickel cyanides complexes are much weaker than copper cyanide complexes, and these metals predominantly exist in the free form (uncomplexed) in presence of cyanide. Thus, a considerable decline in the adsorption of zinc and nickel on the bagasse surface (at pH 1.26) could be due to the electrostatic repulsion between the sugarcane bagasse surface (positively charged) and the free zinc and nickel ions (cations).

Synthetic wastewater	Cu(II) (200 mg/L^{-1})	Ni(II) (14.5 mg/L^{-1})	Zn(II) (45.5mg/L^{-1})	CN$^-$ (18.1mg/L^{-1})
	q (mg metal / g adsorbent)			
Initial	1.85	0.145	0.421	_
with cyanide	1.89	0.134	0.412	18.1
After cyanide removal	2.04	0.144	0.443	ND

Table 8. Adsorption capacities of toxic metal ions from simulated electroplating wastewater using sugar cane bagasse. Conditions: pH 1.26; flow: 2.0 mL/min; mass of the adsorbent: 4.0g; temperature: 28 ±2ºC.

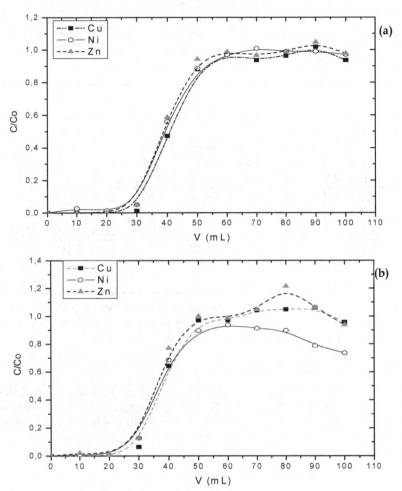

Fig. 11. Breakthrough curves of metals. (a) wastewater with cyanide (18 mg/L) and (b) wastewater after NaOCl oxidation of cyanide. Amount of adsorbent 4.0g, Cu^{+2} concentration 200 mg/L at pH 1.78 and flow rate of 2.0mL/min. C/C_o represent the ion concentrations for initial and final solution. (Adaptaded from Sousa et al., 2009. Journal Environment Manager, Vol. 90, N° 11, pp. 3340–3344, ISSN 0301-4797)

3.3.3 Column application in samples industrial

Electroplating wastewater contains metal ions, then an effective removal of these species is necessary. For testing the treatability from an wastewater industrial (Juazeiro do Norte – Ceara -Brazil) containing Cu- Ni-Zn with sugar cane bagasse, thus 250 mL wastewater were passed on column in conditions as described previously to obtain the breakthrough. 60 mL of the final solution for both metals, at pH 1.5, wcre sufficient to saturate the amount of adsorbent (4.0 g) on column. Table 9 shows the metals ions removal (Cu, Ni and Zn) from an electroplating-wastewater industrial (Juazeiro do Norte – Ceara -Brazil) by sugar cane bagasse. For testing the treatability from an wastewater industrial was employed a polyethylene column (10cm x 0.8 cm D.I) packed with 0.30g of adsorbent (bed depth of 6 cm) at flow rate 1.0 mL/min[-1] at pH 5. It is observed in Table 9 that the copper removal varied 42.4 to 90.8% and 13.7 to 52.6% for nickel from wastewaters.

Metal ion	Conc.	Wastewater samples							
		1	2	3	4	5	6	7	8
Cu	C_0 (mg L[-1])	73.03	17.44	147.03	3.95	17.84	0.46	0.43	40.99
	C_e (mg L[-1])	19.77	1.60	36.75	0.41	2.02	-	0.099	5.39
	Removal,%	72.92	90.80	75.00	89.70	88.69	-	77.35	86.84
Ni	C_0 (mg L[-1])	38.45	1.15	17.94	1.52	1.62	2.28	1.58	0.82
	C_e (mg L[-1])	18.98	0.77	9.09	0.77	0.91	1.34	0.75	-
	Removal,%	50.64	33.21	49.33	49.62	43.96	41.14	52.59	-
Zn	C_0 (mg L[-1])	40.13	10.53	31.29	-	-	-	-	-
	C_e (mg L[-1])	7.15	3.34	7.26	-	-	-	-	-
	Removal,%	82.19	68.11	76.80	-	-	-	-	-

Table 9. Metals removal from industrial wastewaters collected from galvanic industries (Juazeiro do Norte- Ceara-Brazil). Conditions: Polyethylene column (10cm x 0.8 cm D.I, packed with 0.30g adsorbent) in the conditions described in previously

3.3.4 Column regeneration and metal recovery

Metal recoveries and the column regeneration were carried out by acid elution method (HCl 0.1M, HNO$_3$ 0.1M), under conditions tested for breakthrough curve, with synthetic solution. The results obtained are shown in Fig.12, 30 mL of acid eluent is efficient for almost complete desorption of the retained metals. Also is noted that the first aliquot of 5 mL elutes more 70 % of retained ions on column, while the remaining eluant volume (25 mL) desorbed the rest

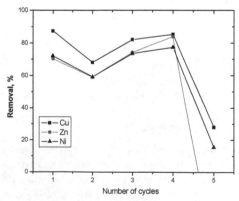

Fig. 12. Removal efficiency with number of cycles for metal ions on cane sugar bagasse from wastewater ((Adaptaded from Sousa et al., 2009. Journal Environment Manager, Vol. 90, N° 11, pp. 3340–3344, ISSN 0301-4797)

of ions. A test for check the recycling of the adsorbent on column was carried out to passing first water (30 mL) on column, and 60 mL of sample solution without treatment, and HCl 2M (70mL) to the elution of metals retained, this procedure was repeated 5 times with the same adsorbent of the column. The results shown in Fig.11, indicate that the removal efficiency decreases from 79.6% to 65 for (Cu), 77.4 to 66.7 % for (Zn) and 73,8 % for (Ni) during their first to fourth cycles. In contrast, a lost drastically of efficiency of the adsorbent was observed after 4 cycles.

4. References

Abdel-Halim, E. S., Abou-Okeil, A. & Hashem, A. (2006). Adsorption of Cr (VI) oxyanions onto modified wood pulp. *Polymer - Plastics Technology and Engineering* . vol. 45 No 1,pp. 71–76, ISSN 1525-6111

Abou-Elela, S. I., Ibrahim, H. S. , Abou-Taleb, E. (2008). Heavy metal removal and cyanide destruction in the metal plating industry: an integrated approach from Egypt. *Environmentalist* vol. 28 No 3,pp. 223-229, ISSN 1573-2991

Agrawal, A., Sahu, K.K. (2009). An overview of the recovery of acid from spent acidic solutions from steel and electroplating industries, Journal of Hazardous Materials Vol. 171,No 1-3, pp.61–75

Akcil, A. (2003). Destruction of cyanide in gold mill effluents: biological versus chemical treatments. *Biotechnology Advances*, Volume 21, No 6,pp. 501-511, ISSN 0734-9750

Akcil, A. (2003). Destruction of cyanide in gold mill effluents: biological versus chemical treatments, Biotechnology Advances, Vol. 21, N° 6, pp. 501-511, ISSN 0734-9750

Aksu, Z., Acikel, U., Kabasakal, E., Treser, S. (2002) Equilibrium modelling of individual and simultaneous biosorption of chromium(VI) and nickel(II) onto dried activated sludge, *Water Research* Vol. 36 No 12 (2002) 3063-3073

Alonso-González, O.;Nava-Alonso, F., Uribe-Salas, A., Dreisinger, D. (2010) Use of quaternary ammonium salts to remove copper–cyanide complexes by solvent extraction, Minerals Engineering, Vol. 23, N° 10, pp. 765-770, ISSN 0892-6875

Amarasinghe,B.M.W.P.K. Williams, R.A. (2007).Tea waste as a low cost adsorbent for the removal of Cu and Pb from wastewater, Chemical Engineering Journal, Vol.132, No 1-3, pp. 299-309, ISSN 1385-8947 .

Amuda, O. S., Giwa, A.A., Bello, I.A.(2007). Removal of heavy metal from industrial wastewater using modified activated coconut shell carbon Biochemical Engineering Journal 2007;Vol. 36 No 2,pp. 174-181, ISSN 1369-703X

Atkinson, BW., Bux F, Kasan HC. 1998. Considerations for application of biosorption technology to remediate metal-contaminated industrial effluents. *Water AS*. Vol. 24, No 2, pp. 129-135. ISSN: 0378-4738

Ayranci, E., Hoda, N.(2005).Adsorption kinetics and isotherms of pesticides onto activated carbon-cloth,*Chemosphere* Vol. 60, No11, pp. 1600-1607, ISSN 0045-6535 60 (2005) 1600–1607.

Babel, S., Kurniawan, A. (2003). Low-cost adsorbents for heavy metal uptake from contaminated water: a review. *Journal of Hazardous Materials*, Vol. 97, No 1-3, pp. 219-243, ISSN 0304-3894

Barakat, M.A, Chen, Y.T, Huang, C.P.(2004). Removal of toxic cyanide and Cu(II) Ions from water by illuminated TiO2 catalyst, Applied Catalysis B: Environmental, Vol. 53, No 1, pp. 13-20, ISSN 0926-3373.

Barreto, A.C.H.; Costa, M.M.; Sombra, A.S.B.; Rosa, D.S.; Nascimento, R.F.; Mazzetto, S.E. & P. B. A. Fechine. (2010). Chemically Modified Banana Fiber: Structure, Dielectrical Properties and Biodegradability. Journal of Polymers and the Environment, Vol. 18, pp 523–531, ISSN 1572-8900

Barreto, A.C.H.; Esmeraldo M.A.; Rosa D.S.; Fechine P.B.A & Mazzetto S.E. (2010). Cardanol biocomposites reinforced with jute fiber: microstructure, biodegradability and mechanical properties, Polymer Composites, Vol. 31, N° 11, pp. 1928–37, ISSN 0272-8397

Barreto, A.C.H.; Rosa, D. S.; Fechine, P. B. A. & Mazzetto, S .E. (2011). Properties of sisal fibers treated by alkali solution and their application into cardanol-based biocomposites. Composites: Part A, Vol. 42, pp. 492–500, ISSN 1359-835X

Behnamfard, A., Salarirad, M.M. (2009). Equilibrium and kinetic studies on free cyanide adsorption from aqueous solution by activated carbon, Journal of Hazardous Materials, Vol. 170, No 1, pp. 127-133, ISSN 0304-3894

Bilba, K. & Ouensanga, A. (1996). Fourier transform infrared spectroscopic study of thermal degradation of sugar cane bagasse. Journal of an analytical And Applied Pyrolysis. Vol. 38, pp. 61-73, ISSN 0165-2370

Bose, P., Bose, MA., Kumar, S. (2002). Critical evaluation of treatment strategies involving adsorption and chelation for waster containing copper, zinc and cyanide. Advances in Environmental Research, Vol. 7, No 1, pp. 179-195, ISSN 1093-0191

Botz, M., Devuyst, E, Mudder, T, Norcross,R., Ou, B.,. Richins, R ,et al. (1995), Cyanide: an overview of cyanide treatment and recovery methods. Mining Environmental Management, Vol. 3, No 2. pp. 4–16. ISSN: 0969-4218

Bower, D.I. & Maddams, W.F. (2006). The vibrational spectroscopy of polymers, Cambridge University Press, ISBN 0521421950, New York.

Brower, J.B., Ryan, R.L. ,Pazirandeh, M. (1997). Comparison of ion-exchange resins and biosorbents for the removal of heavy metals from plating factory wastewater, Environmental Science and Technology Vol. 31,No 10, pp. 2910–2914. ISSN, 1520-5851

Brundle, R.; Evans Jr., C.A. & Wilson S. (1992). Encyclopedia of materials characterization: surfaces, interfaces, thin films, Buttertworth-Heinemann, ISBN 0750691689, London.

Buckeridge, M.S. & Goldman, G.H. (2011) Routes to Cellulosic Ethanol, Springer, ISBN 0387927395, New York.

Butter, TJ., Evison, L.M Hancock, I.C. , Holland, F.S., Matis, K.A., Philipson, A.,. Sheikh, A.I., Zouboulis, A.I. (1998). The removal and recovery of cadmium from dilute aqueous solutions by biosorption and electrolysis at laboratory scale. Water Research. Vol. 32,No 2 ,pp. 400-406 ISSN 0043-1354

Calado V.; Barreto D.W. & D´Almeida, J.R.M. (2000). The Effect of a Chemical Treatment on the Structure and Morphology of Coir Fibers. Journal of Materials Science Letters, Vol. 19, N° 23, pp. 2151-2153, ISSN 0261-8028

Cay, S., Uyanik, A., Ozasik, A.(2004) Single and binary component adsorption of copper(II) and cadmium(II) from aqueous solutions using tea-industry waste, Separation and Purification Technology, Vol. 38, No 3,pp. 273-280, ISSN 1383-5866

Chang, M.M.; Chou, T.C. & Tsao, G.T. (1981). Structure, Pretreatment and Hydrolysis of Cellulose, In: Advances in Biochemical Engineering/Biotechnology. pp.15-42. Spinger, ISBN 3-540-11018-6, New York.

Clesceri, L.S., A.E. Greenberg, and A.D.Eaton, (Editors). (1998). *Standard Methods for the Examination of Water and Wastewater*. 20th Edition. ISBN: 0875532357. American Public Health Association. Washington, D.C. 1325 p.

Cooney, D.O.(1999). *Adsorption design for wastewater treatment* , Lewis publishers, ISBN 1566703336 9781566703338. Boca Raton, Fl

Crini, G. (2005). Recent developments in polysaccharide-based materials used as adsorbent in wastewater treatment. *Progress in Polymer Science*. Vol.30, No 1, 38-70, ISSN 0079-6700.

Crini, G. (2006). Non-conventional low-cost adsorbents for dye removal: a review. *Bioresource Technology* Vol. 97,No 9,pp. 1061-1085,. , ISSN 0960-8524

Crisafully, R.; Milhome, M.A; Cavalcante, R.M., Silveira, E.R; Nascimento, R.F. (2005) Adsorption of PAHs from petrochemical plant wastewater using low cost adsorbents. *Separation and purification Technology* , Vol. 99, No 10, pp. 4515-4519, ISSN 0960-8524

Dąbrowski, D., (2001) Adsorption — from theory to practice, *Advances in Colloid and Interface Science*, Vol. 93, No 1-3,pp. 135-224, ISSN 0001-868

Dash, R., Balomajumder, C. & Kumar, A. (2008). Removal of cyanide from water and wastewater using granular activated carbon. *Chemical Engineering Journal*, Vol 146, No 3, pp. 408-413, ISSN 1385-8947

Deosarkar S.P, Pangarkar V. G. Adsorptive separation and recovery of organics from PHBA and SA plant effluents.(2004),*Separation and Purification Technology*, Vol. 38, No 3, pp. 241-254, ISSN 1383-5866

Duncan, J.R., Brady,D., Stoll, A. (1994). Biosorption of Heavy Metals Cations by Non-Viable Yeast Biomass *Environmental Technology*, Vol.15 No 5 pp. 429-438,ISSN: 1479-487X

Dutra, A. J. B., Rocha, G. P. & Pombo, F. R. (2008). Copper recovery and cyanide oxidation by electrowinning from a spent copper-cyanide electroplating electrolyte. *Journal of Hazardous materials* Vol.. 152, No 21, pp. 648–655, ISSN 0304-3894

Ebbs, S. (2004). Biological degradation of cyanide compounds, *Current Opinion in Biotechnology* Vol.15,No 3 pp. 231-236 ISSN 0958-1669.

Esmeraldo, M.A. (2006). Preparação de Novos Compósitos Suportados em Matriz de Fibra Vegetal, Master's Degree, Departamento de Química Orgânica e Inorgânica, Universidade Federal do Ceará, Fortaleza-CE-Brazil.

Esmeraldo, M.A.; Barreto, A.C.H.; Freitas, J.E.B.; Fechine, P.B.A.; Sombra, A.S.B.; Corradini, E.; Mele, G.; Maffezzoli, A. & Mazzetto, S.E. (2010). Dwarf-green coconut fibers: a versatile natural renewable raw bioresource. *Treatment morphology and physicochemical properties*. *Bioresources*, Vol. 5, N° 4, pp. 2478–2501, ISSN 1930-2126

Febrianto, J., Kosasih, A.N., Sunarso, J., Ju, Y-H., Indraswati, N., Ismadji, S. (2009) Equilibrium and kinetic studies in adsorption of heavy metals using biosorbent: A summary of recent studies, *Journal of Hazardous Materials*, Vol. 162, No 2-3, , pp. 616-645, ISSN 0304-3894.

Fernando, K., Lucien, F., Tran T., Carter, M. (2008). Ion exchange resins for the treatment of cyanidation tailings. Part 3. Resin deterioration under oxidative acid conditions. *Minerals Engineering*. Vol. 21, No 10,pp. 683–690, ISSN 0892-6875

Franson MH. (1992). *Standard methods for the examination of water and wastewater*. In: Franson, editor, 18th ed. Washington, DC: American Public Health Association, American Water Works Association, Water Environment Federation.

Freeman, H.M. (1989). *Standard Handbook of Hazardous Waste Treatment and Disposal*,(Ed.) McGraw-Hill, New York, New York, NY, ISBN 0-07022042-5.

Freundlich, H.M.F. (1906). Uber die adsorption in lösungen, Zeitschrift für Physikalische Chemie - 470, vol.57(A), pp 385 ISSN 0942-9352

Georgopoulos, S.T.; Tarantili, P.A.; Avgerinos E.; Andreopoulos A.G. & Koukios, E.G. (2005). Thermoplastic polymers reinforced with fibrous agricultural residues. *Polymer Degradation and Stability*, Vol. 90, pp. 303–312, ISSN 0141-3910

Gijzen, H.J. (2000). Cyanide toxicity and cyanide degradation in anaerobic wastewater treatment. *Water Research* Vol. 34,No 9, pp. 2447, ISSN 0043-1354

Gilbert, R.D. (1994). Cellulosic polymers, blends and composites, 115–130. *Hanser Publishers*, ISBN 3-446-16521-5, New York.

Guo, X., Zhang, S. & Shan, X. (2008). Asorption of metal ions on lignin. *Journal of Hazardous Materials* Vol., Vol. 151,No 1,pp. 134–142, ISSN 0304-3894

Gupta, B. S., Curran A, M., Hasan, S. & Ghosh, T. K. (2009). Adsorption characteristics of Cu and Ni on Irish peat moss. *Journal of Environmental Management* Vol.90,No 2,pp. 954-960, , ISSN 0301-4797

Gupta, V.K., Srivastava, S.K. Mohan, D., Sharma, S. (1997). Design parameters for fixed bed reactors of activated carbon developed from fertilizer waste for the removal of some heavy metal ions, Waste Management, Vol 17, Nº 8, 1997, pp. 517-522, ISSN 0956-053X.

Han, B., Shen, Z., Wickramasinghe, S. (2005). Cyanide removal from industrial wastewaters using gas membranes. *Journal of Membrane Science Vol.* 257, No 1-2, pp. 171–181, ISSN 0376-7388

Han, S. O. & Choi, H. Y. (2010). Morphology and surface properties of natural fiber treated with electron beam, Microscopy: *Science, Technology, Applications and Education*, Vol. 3, ISBN 978-84-614-6191-2, pp. 1880-1887

Hashem, A. (2006). Amidoximated sunflower stalks (ASFS) as a new adsorbent for removal of Cu (II) from aqueous solution. *Polymer - Plastics Technology and Engineering* 45, No 1,pp. 35–42, ISSN 1525-6111

Hashem, A., Abdel-Halim, E.S., El-Tahlawy, K H. F. & Hebeish, A. (2005). Enhancement of adsorption of Co (II) and Ni (II) ions onto peanut hulls though esterification using citric acid. *Adsorption Science and Technology* Vol.23, No 5 pp. 367–380, ISSN- 0263-6174

Hashem, A., Abou-Okeil, A., El-Shafie, A. & El-Sakhawy, M. (2006). Grafting of high α-cellulose pulp extracted from sunflower stalks for removal of Hg (II) from aqueous solution. *Polymer - Plastics Technology and Engineering* Vol.45,No 1 pp. 135–141 ISSN 1525-6111

Hashem, A., Akasha, R.A., Ghith A., Hussein, D.A. (2007). Adsorbent based on agricultural wastes for heavy metal and dye removal: a review. *Energy Education Science and Technology*. Vol.19, pp. 69–86 ISSN 1301-8361

Hashem, A., Aly, A. A. & Aly, A. S. (2006). Preparation and utilization of cationized sawdust. *Plastics Technology and Engineering* Vol. 45, No. 3,pp. 395–401, ISSN 1525-6111

Hashem, A., Aly, A. A., Aly, A. S. & Hebeish, A. (2006). Quaternization of cotton stalks and palm tree particles for removal of acid dye from aqueous solutions. *Plastics Technology and Engineering* Vol.45,No 3, pp. 389–394, ISSN 1525-6111

Hashem, A., Elhammali M. M., Hussein, A. H. & Senousi, M. A. (2006). Utilization of sawdust-based materials as adsorbent for wastewater treatment. *Plastics Technology and Engineering*.Vol. 45,No 7, pp. 821–827, , ISSN 1525-6111

Hashem, A., Elhmmali, M. M., Ghith, A., Saad, E.E. & Khouda, M. M. (2007). Utilization of chemically modified Alhagi residues for the removal of Pb (II) from aqueous solution. *Energy Education Science and Technology* 20, pp. 1–19, ISSN 1301-8361

Hayashi, A. M. (2001). Remoção de Cromo Hexavalente através de Processo de Biossorção em Algas Marinhas. *Tese de Doutorado da Faculdade de Engenharia Química*, Unicamp, Campinas, SP, p 20, 22, 61, 62 –63, 80, 82 –83, 86

Ho, YS, McKay, G. The kinetics of sorption of basic dyes from aqueous solutions by sphagnum moss peat(1998). Canadian Journal of Chemical Engineering,Vol.76, No 4,pp. 822-827. ISSN 1939-019X

Igwe, J.C., Abia, A.A. (2007). Adsorption kinetics and intraparticulate diffusivities for bioremediation of Co(II), Fe(II) and Cu(II) ions from waste water using modified and unmodified maize cob. International Journal of the Physical Sciences, vol. 2 5, pp. 119–127. ISSN 1992-1950

Ismail, I., Abdel-Monem, N.; Fateen, S-E., Abdelazeem, W. (2009). Treatment of a synthetic solution of galvanization effluent via the conversion of sodium cyanide into an insoluble safe complex, Journal of Hazardous Materials, Vol. 166, No 2-3, pp. 978-983, ISSN 0304-3894

Isom, G.E., Borowitz, J.L. (1995). Modification of cyanide toxicodynamics : mechanistic based antidote development, *Toxicology Letters* Vol.82-83,pp.795-799, ISSN 0378-4274

Jnr, M.H., Spiff, A. I. (2005). Effects of temperature on the sorption of Pb2+ and Cd2+ from aqueous solution by Caladium bicolor (Wild Cocoyam) biomass, *Electronic Journal of Biotechnology* Vol. 8, No 2,162-169. ISSN 0717-3458

Joint Committee on Powder Diffraction Standards (JCPDS) - International Center for Diffraction Data. (1986). JCPDS File 50-2241.

Kavitha, D.,Namasivayam, C. Experimental and kinetic studies on methylene blue adsorption by coir pith carbon, (2007). *Bioresource Technology*, Vol. 98, No 1, pp. 14-21, ISSN 0960-8524

Klenk, H., Griffiths, A., Huthmacher, K., Itzel, H., Knorre,H., Voight, C., et al. (1996). Cyano, inorganics, in: *Ullmanns encyclopedia of industrial chemistry*, A8 ,in: W Gerhartz, YS Yamamoto, L Kaudy, R Pfefferkorn, JF Rounsaville, Editors VCH, pp. 159–190. ISBN 978-3527303854

Kumar, M.N.V.R. (2000). A review of chitin and chitosan applications. *Reactive and Functional Polymers* Vol.46: No 1, pp. 1-27, ISSN 1381-5148

Kumar, U,, Bandyopadhyay, M. (2006). Sorption of cadmium from aqueous solution using pretreated rice husk, *Bioresource Technology*, Vo. 97, No 1,pp. 104-109, ISSN 0960-8524

Kundu, S. Gupta, A.K. (2005). Analysis and modeling of fixed bed column operations on As(V) removal by adsorption onto iron oxide-coated cement (IOCC), Journal of Colloid and Interface Science, Vol. 290, No 1, pp. 52-60. ISSN: 0021-9797

Kurama, H.,Çatalsarik, T. (2000). Removal of zinc cyanide from a leach solution by an anionic ion-exchange resin, *Desalination*, Vol. 129, No 1,pp. 1-6, ISSN 0011-9164

Lagergren, S. Zur theorie der sogenannten adsorption geloester stoffe,Veternskapsakad Handlingar 24 (1898) 1–39.

Langmuir, I. (1918).The adsorption of gases on plane surfaces of glass, mica and platinum. Journal of the American Chemical Society, Vol 40, N° 9, pp 1361–1403, ISSN: 0002-7863

Lanza, M. & Bertazzoli R. (2002). Cyanide oxidation from wastewater in a flow electrochemical reactor. *Industrial and Engineering Chemistry Research* Vol. 41 No 1, pp. 22–26, ISSN 1520-5045

Liu, F.; Wu, J.; Chen, K. & Xue, D. (2010). Morphology Study by Using Scanning Electron Microscopy, Microscopy: *Science, Technology, Applications and Education*, Vol. 3, ISBN 978-84-614-6191-2, pp. 1781-1792.

López, R.; Gutarra, A.(2000). Descoloração de águas residuais da indústria têxtil. *Química Têxtil*, vol. 59, pp. 66-69, ISSN 0102-8235.

Low, K.L, Lee,C.K, Mak,S.M. (2004). Sorption of copper and lead by citric acid modified Wood, *Wood Science and Technology* Vol.38, No 8, pp.629-640, ISSN 1432-5225

Low, K.S., Lee, C.K. (1991). Cadmium update by moss colympers delesertic, besch, *Bioresource Technology* Vol. 38, No 1 ,pp.1–6. ISSN 0960-8524

Lu, J., Dreisinger, D.B. and Cooper, W.C. (2002). Thermodynamics of the aqueous copper-cyanidesystem. *Hydrometallurgy*, Vol. 66,No 1–3, pp. 23–36, ISSN 0304-386X

Luo, F., Liu, Y.,Li, X., Xuan ,Z. , Ma, J. (2006). Biosorption of lead ion by chemically-modified biomass of marine brown algae Laminaria japonica. *Chemosphere* Vol. 64, No 7,pp. 1122-1127, ISSN 0045-6535

Mattuschka, B., Straube, G., J (1993) Journal of Chemical Technology and Biotechnology,Vol. 58,No 2,pp. 157-63. ISSN: 1097-4660

McKay, G. (1996) (Editors), Design of adsorption contatacting systems,in: Use of Adsorbents for the Rremoval of Pollutans from Wasterwaters,pp 99-131,CRC press,Inc, Boca Raton, Florida ISBN: 0-8493-6920-7

Mohan, D.; Singh, K. P., (2002) Single- and multi-component adsorption of cadmium and zinc using activated carbon derived from bagasse—an agricultural waste, *Water Research*,Vol. 36, No 9,pp. 2304-2318, ISSN 0043-1354

Monser L. & Adhoum, N. (2002). Modified activated carbon for the removal of copper, zinc, chromium and cyanide from wastewater. *Separation and Purification Technology*. Vol. 26,No 2-3, pp. 137–146, ISSN 1383-5866.

Monteagudo, J., Rodríguez, L. ,Villaseñor, J. (2004). Advanced oxidation processes for destruction of cyanide from thermoelectric power station waste waters. *Journal of Chemical Technology and Biotechnology*. Vol. 79 No 2, (June 2011) pp. 117–125, 1097-4660 ISSN 1097-4660

Morán, J.I.; Alvarez, V.A.; Cyras, V. & Vázquez, A. (2008). Extraction of cellulose and preparation of nanocellulose from sisal fibers. *Cellulose*, Vol. 15, pp. 149–159, ISSN0969-0239

Moreira, S. A., Sousa, F. W., Oliveira, A. G., Brito, E.S, and Nascimento, R. F. (2009). "Metal removal from aqueous solution using cashew bagasse, Química Nova vol. 32, No 7, pp.1717-1722. ISSN 0100-4042.

Morgan, S.M. & Lee,C.M. (1997). Metal and acid recovery options for the plating industry, *Resources, Conservation and Recycling* Vol.19,No 1, pp. 55–71, ISSN: 09213449

Moussavi, G., Talebi, S. (2011). Comparing the efficacy of a novel waste-based adsorbent with PAC for the simultaneous removal of chromium (VI) and cyanide from electroplating wastewater, Chemical Engineering Research and Design, ISSN 0263-8762, doi.10.1016/j.cherd.2011.10.014.(*in press*)

Moussavi,, G.; Khosravi, R. (2010). Removal of cyanide from wastewater by adsorption onto pistachio hull wastes: Parametric experiments, kinetics and equilibrium analysis, Journal of Hazardous Materials, Volume 183, N° 1-3, 15,pp 724-730, ISSN 0304-3894,

Mudder, T. & Botz, M. (2004). Cyanide and society: a critical review. *European Journal of Mineral Processing and Environmental Protection* Vol.4, No 1, pp. 62–74, ISSN 1303-0868.

Nagarale, R.K., Gohil, G.S.,. Shahi, V. K. (2006). Recent developments on ion-exchange membranes and electro-membrane processes, *Advances in Colloid and Interface Science* Vol. 119, No 2-3 97-130, ISSN 0001-8686

Naja, G., Murphy, V. & Volesky, B. (2010). Biosorption, Metals, In *Encyclopedia of Industrial Biotechnology: Bioprocess, Bioseparation, and Cell Technology*.pp. 1–29. John Wiley & Sons, DOI: 10.1002/9780470054581.eib166

Neto, V. O. S., Oliveira, A. G., Teixeira, R. N. P., Silva, M. A. A., Freire, P. T. C., Keukeleire, D. D., Nascimento, R. F. (2011). Use of coconut bagasse as alternative adsorbent for separation of copper(II) ions from aqueous solutions: Isotherms, kinetics, and thermodynamic studies, *BioResources* Vol. 6 , No3,pp. 3376-3395. ISSN: 1930-2126

Ngah, W. S. W., Teong, L. C. & Hanafiah, M. A. K. M. (2011). Adsorption of dyes and heavy metal ions by chitosan composites: A review. *Carbohydrate Polymers* Vol.83, No 4-1, pp. 1446-1456, ISSN 0144-8617

Ngah, W.S. W., Hanafiah, M.A.K.M. (2008). Adsorption of copper on rubber (Hevea brasiliensis) leaf powder: Kinetic, equilibrium .and thermodynamic studies, Biochemical Engineering Journal, Vol. 39, N° 3, pp. 521-530, ISSN 1369-703X.

Noeline, B.F; Manohar, D.M.; Anirudhan, T.S. (2005). Kinetic and equilibrium modelling of lead(II) sorption from water and wastewater by polymerized banana stem in a batch reactor, *Separation and Purification Technology*, Vol. 45, No 2, pp.131-140, ISSN 1383-5866

Ouajai, S. & Shanks, R,A, (2005). Composition, structure and thermal degradation of hemp cellulose after chemical treatments. *Polymer Degradation and Stability*. Vol. 89, pp. 327–335, ISSN 0141-3910

Pagnanelli, F., Mainelli, S., Veglio, F. & Toro, L. (2003). Heavy metal removal by olive pomace: biosorbent characterization and equilibrium modeling. *Chemical Engineering Science* Vol. 58, No pp. 4709–4717, ISSN 0009-2509

Park, D., Kim, Y., Lee, D. & Park, J. (2008). Chemical treatment for treating cyanides-containing effluent from biological cokes wastewater treatment process. *Chemical Engineering Journal* Vol.143, No 1-3, pp. 141–146, ISSN 1385-8947

Pasavant, P., Apiratikul, R., Sungkhum, V., Suthiparinyanony, P., Wattanachira, S., Marhaba, T. F., (2006) Biosorption of Cu2+, Cd2+, Pb2+, and Zn2+ using dried marine green macroalga Caulerpa lentillifera *Bioresource Technology* Vol.97 , No 18 pp. 2321-2329. , ISSN 0960-852

Patil, Y.B, Paknikar, K.M. (2000). Development of a process for biodetoxification of metal cyanides from wastewater, *Process Biochem*. Vol.35, No 10, pp. 1139–1151, ISSN 1359-5113

Patil, YB., Paknikar, KM. 1999. Removal and recovery of metal cyanides using a combination of biosorption and biodegradation processes. *Biotechnology Letters*, Vol. 21, No 10,pp.913-919. ISSN: 1573-6776

Qiu, T. S., Cheng, X. X., Hao, Z. W. & Lou, X. P. (2002). Present situation and development for wastewater containing cadmium treatment technology. [J]. *Sichuan Nonferrous Metals*, No 4: pp. 38-41 ISSN 1006-4079

Regel-Rosocka, M. (2010). A review on methods of regeneration of spent pickling solutions from steel processing, *Journal of Hazardous Materials* Vol. 177, No 1-3, 57–69, ISSN 0304-3894

Saarela, K., Kuokkanen, T., Peramaki, P. & Valimaki, I. (2005). Analyses and treatment methods of waste water containing metal cyanides. *Journal of Solid Waste Technology and Management vol.* 31, No 1, pp. 38–45, ISSN: 1088-1697

Saikaew, W., Kaewsarn,P., Saikaew, W. (2009). Pomelo Peel: agricultural waste for biosorption of cadmium ions from aqueous solutions, World Academy of Science, *Engineering and Technology*. Vol.56, pp.287–291. INSS 56 287–291

Sankararamakrishnan, N. , Sharma,A.K., Sanghi,R., Novel chitosan derivative for the removal of cadmium in the presence of cyanide from electroplating wastewater(2007), Journal of Hazardous Materials, Vol. 148, No 1-2, ,pp. 353-359, ISSN 0304-3894

Sekhar, K. C.; Kamala, C. T.; Chary, N. S.; Anjaneyulu, Y., Inter. J. Min. Processing, 68(2003)

Şengil, İ.A., Özacar, M., Türkmenler, H. (2009). Kinetic and isotherm studies of Cu(II) biosorption onto valonia tannin resin, *Journal of Hazardous Materials*, Vol. 162, No 2-3,pp. 1046-1052, ISSN 0304-3894

Seung-Mok, Lee; Diwakar, Tiwari. (2009). Application of ferrate(VI) in the treatment of industrial wastes containing metal-complexed cyanides: A green treatment, Journal of Environmental Sciences, V. 21, Nº 10, pp 1347-1352, ISSN 1001-0742,

Shen, J. & Duvnjak, Z. (2005). Adsorption kinetics of cupric and cadmium ions on corncob particles. *Process Biochemistry* Vol. 40,No 11, pp. 3446-3454, ISSN 1359-5113

Shibi, I.G; Anirudhan T.S. (2005). Adsorption of Co(II) by a carboxylate-functionalized polyacrylamide grafted lignocellulosics, *Chemosphere* vol. 58, No 8,pp. 1117–1126, ISSN 0045-6535

Silva, R.M.P., Rodríguez,A. B., De Oca, J.M.G., Moreno, D.C. (2009). Biosorption of chromium, copper, manganese and zinc by Pseudomonas aeruginosa AT18 isolated from a site contaminated with petroleum. *Bioresource Technology*, Vol. 100,No 4, pp. 1533–1538, ISSN 0960-8524

Singh, K., Mohan, S. (2004). Kinetic studies of the sucrose adsorption onto an alumina interface, Applied Surface Science, Vol 221, No 1-4, pp. 308-318, ISSN 0169-4332.

Smith, A., Mudder, T. (1991). The Chemistry and Treatment of Cyanidation Wastes,(2th Edition), *Mining Journal Books Limited*, pp. 327-333, ISBN: 0900117168, London

Solomonson, L.P. (1981). Cyanide as a metabolic inhibitor. In: *Cyanide in Biology* , B. Vennesland, E.E. Conn, C.J. Knowles, J. Westley and F. Wissing, Editors, pp. 385-314, Academic Press, 0127169806, London.

Sousa, F. W. (2007). Adsorção de metais tóxicos em efluente aquoso usando pó da casca de coco verde tratado. *Dissertação de Mestrado*, Universidade Federal do Ceará. Departamento de Engenharia Hidráulica e Ambiental, Fortaleza, Ceará, Brasil

Sousa, F. W., Moreira, S. A., Oliveira, A. G., Cavalcante, R. M., Rosa, M. F, and Nascimento, R. F. (2007). The use of green coconut shells as absorbents in the toxic metals, Química. Nova. vol.30, n.5, pp. 1153-1157. ISSN 0100-4042.

Sousa, F. W.,Oliveira, A.G., Ribeiro, J.P., De Keukeleire, Denis., Sousa, A.F., Nascimento, R.F. (2011).Single and multielementary isotherms of toxic metals in aqueous solution using treated coconut shell powder. Desalination and Water Treatment, Vol. 36, pp. 289-296, ISSN 1944-3994

Sousa, F. W.;Oliveira, A. G. ; Ribeiro, J. P.; Keukeleire, D.; Nascimento, R.F.; Rosa, M. F. (2010). Green coconut shells applied as adsorbent for removal of toxic metal ions using fixed-bed column technology. Journal of Environmental Management, vol. 91, pp. 1634-1640. ISSN 0301-4797

Sousa, F.W.; Sousa, M.J.; Oliveira, I.R.N.; Oliveira, A.G.; Cavalcante, R.M.; Fechine, P.B.A.; Neto, V.O.S.; Keukeleire, D. & Nascimento, R. F. (2009). Evaluation of a low-cost adsorbent for removal of toxic metal ions from wastewater of an electroplating factory. Journal Environment Manager, Vol. 90, N° 11, pp. 3340-3344, ISSN 0301-4797

Stuart, B.H. (2004). *Infrared Spectroscopy: Fundamentals and Applications*. ANTS – John Wiley & Sons, ISBN 978-0-470-85428-0, San Francisco.

Szpyrkowicz, L.; Kaul, S.N; Molga, E.; DeFaveri, E.; (2000). Comparison of the performance of a reactor equipped with a Ti/Pt and an SS anode for simultaneous cyanide removal and copper recovery, Electrochimica Acta, Vol. 46, N° 2-3, pp 381-387, ISSN 0013-4686,

Torma, A.E. e Apel, W.A. (1991). Recovery of Metals from Dilute Effluent Streams by Biosorption Methods. In: R.G. Reddy, W.P. Imrie e P.B. Queneau (Eds.,Residues and Effluents - *Processing and Environmental Consideration*. TMS, p.735

Ulbricht, M. (2006). Advanced functional polymer membranes, Polymer Vol. 47 No 7, pp. 2217-2262. ISSN 0032-3861.

Vargas, T., Jerez, C.A., Wierts, J.V.,Toledo, H.(Eds). (1995). *Biohydrometallurgical Processing: Vol. I: Microbiology and Dissolution Mechanisms in Bioleaching: Bioleaching Processes for Gold, Copper and Non-Sulfide Ores*, Universidad de Chile, Viña del Mar, ISBN: 956-19-0209-5.

Vázquez, G.; Álvarez, J. G.; Freire, S.; Lorenzo, M. L.; Antorrena, G., (2002) Removal of cadmium and mercury ions from aqueous solution by sorption on treated Pinus pinaster bark: kinetics and isotherms, *Bioresource Technology*, Vol. 82, No 3, pp. 247-251, ISSN 0960-8524

Viera, R.G.P.; Filho, G. R.; de Assunção, R.M.N.; Meireles, C. S.; Vieira, J. G. & Oliveira, G. S. (2007). Synthesis and characterization of methylcellulose from sugar cane bagasse cellulose. *Carbohydrate Polymers*, Vol. 67, pp. 182-189, ISSN 0144-8617

Vijayaraghavan, K., Yun, Y-S. (2008). Bacterial biosorbents and biosorption, *Biotechnology Advances*, Vol. 26, No 3,pp. 266-291, ISSN 0734-9750

Vilar, V.J.P. , Botelho, C.M.S., Boaventura, R.A.R. (2007). Chromium and zinc uptake by algae Gelidium and agar extraction algal waste: Kinetics and equilibrium. Journal of Hazardous Materials Vol. 149, No 3 643-649, ISSN 0304-3894.

Volesky, B.(Ed.). (1990). *Biossorption of Heavy Metals*.: CRC Press, Inc., ISBN-10: 0849349176, Boston, USA

Waalkes, M.P. (2000),Cadmium carcinogenesis in review. *Journal of Inorganic Biochemistry.* Vol. 79 No 1-4,pp. 241-244, ISSN 0162-0134

Wang, P., Min, X. B. & Chai, L. Y. (2006). The status of treatment technology on wastewater containing cadmium and the development of its bio-treatment technology. [J]. *Industrial Safety and Dust Control,,* Vol. 32, No 8,pp. 14-16 ISSN- 1001-425X

Watanabe, M., Kawahara,K., Sasaki, K. & Noparatnaraporn, N. (2003). Biosorption of cadmium ions using a photosynthetic bacterium, Rhodobacter sphaeroides S and a marine photosynthetic bacterium, Rhodovulum sp. and their biosorption kinetics. *Journal of Bioscience and Bioengineering* 95, pp.374-378 ISSN 1347-4421

Weber, W.J & Morris, J.C. (1963). Kinetics of adsorption on carbon from solution (1963), Journal Sanitary Engineering Division. Proceedings American Society of Civil Engineers. Vol. 89 pp.31-59. ISBN/ISSN: 0044-7986.

White, C. , Wilkinson, S.C., Gadd, G.M. (1995). The role of microorganisms in biosorption of toxic metals and radionuclides, *International Biodeterioration & Biodegradation*, Vol. 35, No 1-3(August 2011),pp. 17-40, ISSN 0964-8305

Yalçin, S., Apak., R, Hizal J., Afsr, H. 2001. Recovery of copper(II) and chromium (III, VI) from electroplating-industry wastewater by ion exchange. *Separation Science and Technology*, Vol.36, No 10, pp. 2181-2196. ISSN 1520-5754

Yang, H.; Yan, R.; Chen, H.; Lee, D.H.; Zheng, C. (2007). Characteristics of hemicellulose, cellulose and lignin pyrolysis. *Fuel.* Vol. 86, pp. 1781–1788, ISSN 0016-2361

Yasemin, B. & Zeki, T. (2007). Removal of heavy metals from aqueous solution by sawdust adsorption. *Journal of Envirnnmental Sciences* Vol. 19,No 2, pp. 160-166, ISSN: 1001-0742

Yazici, E.Y., Deveci, H., Alp I. & Uslu, T. (2007). Generation of hydrogen peroxide and removal of cyanide from solutions using ultrasonic waves. *Desalination* Vol. 216, No 1-3, pp. 209–221, ISSN 0011-9164

Zafar, M.N., Nadeem, R.,Hanif, M.A. (2007). Biosorption of nickel from protonated rice bran, *Journal of Hazardous Materials* Vol.143, No 1-2, 478–485, ISSN 0304-3894

Zvinowanda, C., Okonkwo J. , Gurira, R. (2008). Improved derivatisation methods for the determination of free cyanide and cyanate in mine effluent. *Journal of Hazardous Materials*, Vol. 158, No 1, pp. 196–201,ISSN 0304-3894

Part 3

Special Topic

7

Experimental Design and Response Surface Analysis as Available Tools for Statistical Modeling and Optimization of Electrodeposition Processes

Lilian F. Senna and Aderval S. Luna
State University of Rio de Janeiro, Chemistry Institute,
Pavilhão Haroldo Lisboa da Cunha, Rio de Janeiro,
Brazil

1. Introduction

Electrodeposition, also called electroplating, is one of the most commonly used methods for metal and metallic-alloy film preparation in many technological processes. This method is generally considered an economically interesting and easily controlled process to protect and enhance the functionality of parts used in diverse industries (home appliances, jewelry, automotive, aircraft/aerospace, and electronics) in both decorative and engineering applications. It promotes the appearance, extends the life, and improves the performance of materials and products in different media (Schwartz, 1994; Oriňáková et al., 2006).

Despite these applications, electrodeposition methods have, for decades, been industrially regarded as a means for the mass production of cheap materials and not for the production of advanced materials with high values. Their use has been limited to surface protection or plating of decorative metallic layers (Yoshida et al., 2009). The evolution of modern technology, however, has created increased demands on the structures and properties of deposits and directed the emphasis of electrodeposition mainly toward engineering processes and materials technology. Electrodeposition is now regarded as a state-of-the-art technology. For example, electrodeposited Cu is the material of choice for the interconnects in ultra-large-scale integrated (ULSI) circuits, and electrodeposited soft magnetic alloys are an important component of magnetic recording heads (Schwarzacher, 2004; Yoshida et al., 2009). In fact, the deposition of alloy coatings is one of the most rapidly expanding topics in the current literature (Oriňáková et al., 2006; Santana et al., 2006, 2007; Dubent, 2007; Ferreira et al., 2007; Silva et al., 2008; Brankovic et al., 2009; Gupta & Podlaha, 2010, Düzgün et al., 2010). Compared to the electrodeposition of a single metal, alloy coatings show better properties because their chemical composition can be varied to the required function. Alloy coatings are denser, harder, generally more resistant to corrosion, possess better magnetic properties and are suitable for subsequent coating by electrodeposition (Senna et al., 2003; Santana et al., 2006).

Nanostructured coatings, with their advanced properties, can also be produced by electrodeposition, where the nanoparticles are directly attached to the substrate.

Electrodeposition is considered a simple, fast, and inexpensive method and is among the most familiar binder-free techniques employed for the preparation of nanoparticles. In comparison to other techniques, the particle size, crystallographic orientation, mass, thickness, and morphology of the nanostructured materials can be controlled by adjusting the operating conditions and bath chemistry. These nanostructured coatings offer great potential for various applications because of their superior characteristics relative to those of conventional coatings (Mohanti, 2011; Lu & Tanaka, 1996; Huang. & Yang, 2005; Finot, et al., 1999; Gurrappa & Binder, 2008). In addition, recent research has been focused on demonstrating the feasibility of the co-deposition of ceramic materials with metals and polymers to create opportunities for the preparation of novel hybrid nanomaterials and nanostructures. Nanohybrid coatings result from the development of novel electrochemical strategies. These coatings exhibit advantageous properties compared with those of individual materials. The coatings also exhibit properties that cannot be obtained by other methods or with single-phase materials (Gurrappa & Binder, 2008; Tsai et al., 2010; Mosavat et al., 2011).

However, the production of the electrodeposited coatings, especially the alloys and hybrid coatings, is always a complex process. For each application, several critical parameters (chemical and operational) must be accurately controlled to fulfill the required characteristics of the produced coating. The conventional and classical methods of studying a deposition process are usually univariate methods in which most of the deposition parameters are maintained at constant values while one of them is varied in a chosen direction. The use of univariate methods means that, in practice, the best parameter values are often chosen empirically, and the coatings are then produced at "optimum conditions," which may not represent the best conditions to obtain the desired coating properties (Ferreira et al., 2007). In addition, these experiments are time consuming, requiring a large number of experiments, and also produce substantial amounts of wastewater. Moreover, the influence of the combined parameters on the studied variables is not evaluated, which confirms that the optimum levels determined by these methods are generally unreliable (San Martín et al., 1998; Rabiot et al., 1998; Santana et al., 2007a, Santana et al., 2007b). Therefore, to ensure greater reproducibility and quality, the development of a more scientific approach that leads to a better understanding of alloy deposition phenomena is important.

One way to achieve a better approach, given the complexity of the deposition processes, is to use experimental design, response surface methodology, and other statistical techniques, in which all the parameters are varied simultaneously, showing the responses of their synergic and antagonistic interactions. These methodologies both drastically reduce the number experiments needed to optimize the process and give statistical inference on the optimum deposition conditions (Rabiot et al., 1998). These methodologies also allow improvements in both process performance and reliability, thereby decreasing the cost and the volume of produced wastewater in addition to leading to the creation of new coatings systems that can fulfill industrial needs (Ferreira et al., 2007; Santana et al., 2007a). Nevertheless, these tools are not extensively used to enhance the quality of the proposed electrodeposition baths or to achieve realistic optimized conditions, and the reports found in the literature are still related to few research groups. This work presents a review concerning the statistical approach for performing one or multi-responses experiments and its application to the electrodeposition processes of metal and alloy coatings. Therefore, the main topics covered by this review will be the following:

- An overview of the electrodeposition process and the conventional methods used in the literature to perform a study concerning the production of metallic, alloy and hybrid coatings;
- A screening concerning factorial experimental designs and how these designs can be used in electrodeposition processes;
- The surface response method for one response variable or for multi-response variables;
- The optimization of the electrodeposition process using a statistical approach;
- The main results presented in the literature concerning the above-mentioned statistical topics applied to the electrodeposition of metal, hybrid, and alloy coatings.

2. Conventional and classic methods for the study and optimization of electroplating conditions

The production of coatings using electrodeposition was developed in the 1800s and is still considered a simple, low-cost and easy operational process. At a minimum, the electrodeposition of a single metallic coating requires an electrolyte containing the reducible metal ion, a conductive substrate, a counter electrode, a power supply, and a container to hold the electrolyte and electrodes. This simplicity accounts for the appeal of electrodeposition, but may also lead to some basic controls that are needed to ensure reproducibility being neglected (Jeerage & Schwartz, 2004). Different substrate preparations, stirring speeds, applied current densities (or potentials), and deposition temperatures, for example, are some parameters that must be well controlled to produce a coating with the desired qualities and to guarantee reproducibility. When an alloy or a hybrid coating is being produced, or even when a complexant agent or additives are included in the electrolytic bath, the complexity of the electrodeposition process increases, and the control of deposition parameters becomes more critical.

The literature contains several reports that propose electrolytic baths and electrodeposition processes for the production of different kinds of coatings for numerous applications. Figure 1 shows the main parameters that usually need to be controlled in an electrodeposition process. The kind of substrate used, the finish process used to prepare the substrate, the electrolyte composition (with respect to different ions and/or ion concentrations, additives, complexant agents, etc.), the physico-chemical characteristics of the solution (e.g., pH or conductivity), the mode of current used, and other parameters such as temperature or stirring speed can, together or individually, affect the quality and the properties of the produced electrodeposits. In general, most of the reports in the literature have involved the use of conventional univariate methods to study the influence of the deposition parameters on the deposition mechanisms and the coating properties. In these studies, each of the previously discussed parameters has been investigated separately while the others have been held at constant values. The best results obtained under the given conditions were then used to study the effects of another parameter, and so on. Few works include any kind of statistical evaluation, and those that do primarily present the average value and standard deviation. However, it is important to note that the obtained responses of the studied variables are valid only for the chosen experimental conditions. Examples from the literature can be found in later sections of this text.

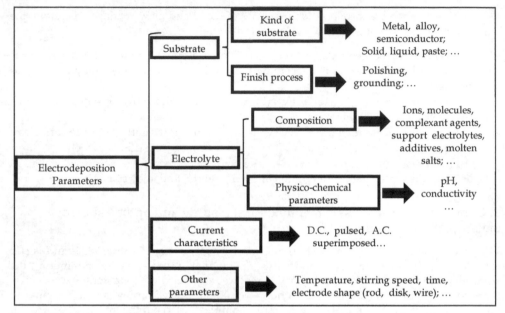

Fig. 1. Main parameters usually controlled in an electrodeposition process.

2.1 Evaluation of the kind of substrate and/or the finish process in the electrodeposition process

Both the kind of substrate used and its finish process may cause differences in the electrodeposition process by influencing the morphology, the growth process, and other properties of the coatings. The substrate and finish process are particularly important when alloys and nanomaterials are produced or when a mechanism is developed. For example, Gomez & Vallés (1999) have noted that the initial stages of the electrodeposition of Co–Ni alloys were influenced by the kind of substrate used. The nucleation and three-dimensional growth, which resulted in compact, fine-grained and homogeneous deposition, occurred in all substrates, with preferential deposition of cobalt and anomalous co-deposition. Concerning the electrodeposition of Zn–Co alloys, Roventi et al. (2006) have observed no significant changes in the polarization curves and the composition of the coatings when the mild steel substrate was changed to platinum. On the contrary, when the substrate tested was glassy carbon, a strong decrease in current density was noted, which started to increase only when the electrode surface was completely covered by the alloy. This behavior was related to a strong inhibition of the deposition process on the glassy carbon cathode.

The properties of the substrate can also influence the coating characteristics. Kim et al. (2009) have shown that the substrate conductivity greatly affected the structural and optical properties of ZnO nanorods produced by cathodic electrodeposition: more conductive substrates favored the growth of high-quality ZnO nanorods at low temperature (80 °C). In addition, studying the effects of Zn substrate finish on the production of ZnO nanoplates, Illy et al. (2005) have observed that the as-received or mechanically polished Zn substrate was the most suitable substrate finish and allowed more highly aligned, dense structures to be grown, whereas electropolished substrates formed inhomogeneous deposits. The investigations concerning this topic are generally performed by changing the substrate and/or its finish treatment while keeping all other parameters constant.

2.2 The effects of other deposition parameters on the coating properties

Because each system substrate/coating may exhibit a different behavior for each electrolyte used, the electrolyte composition, pH and (less often) conductivity are also important and typically studied parameters that can affect the electrodeposition process. Changes in the electrodeposition process occur if only one cation, several cations, or even dispersed particles are present in the electrolyte. Moreover, the solution pH and conductivity, as well as the presence of complexant agents and/or additives, are parameters considered when a new system is under investigation. Other parameters, such as current density, deposition potential, deposition temperature, and stirring speed are also studied. Several works may be used as examples to demonstrate the methodologies usually found in the literature to evaluate these parameters and attempt to find optimum conditions to produce the metallic/alloy/hybrid coatings.

The effects of additives and complexant agents on electrodeposition baths for deposition of metallic and alloy coatings has been investigated by a number of authors. For example, Healy & Pletcher (1998a, 1998b) have shown that the studied additives (Cl-, polyethylene glycol, and 4,5-dithiaoctane-1,8-disulfonic acid) adsorbed on the electrode at different deposition potentials and stirring speeds influences in the copper reduction mechanism. Similar results were observed by Bozzini et al. (2006) and Tantavichet & Pritzker (2006), who studied the effects of polyethylene glycol and benzotriazole, respectively. Nayana et al. (2011) have investigated the influence of additives on the morphology and brightness of nanocrystalline zinc electrodeposited from an acid sulfate bath. The mechanism of zinc deposition also changed in the presence of the additives. Drissi-Daoudi et al. (2003) have studied the influence of several complexant agents on the copper deposition mechanism and showed that the kind of complex formed (strong- or weak-field complex) affected the reduction potential of the metallic ion. The electrodeposition of Cu–Zn alloy on carbon steel from pyrophosphate-base baths in the presence of additives has been investigated by Senna et al. (2003) and Senna et al. (2005). The authors chose one stirring speed and pH value, whereas the bath composition and the current density were varied. Their results have shown that the alloy coating composition, the coating morphology, hardness and corrosion resistance were dependent on these two parameters. Coating compositions similar to that of bulk brass (70 % m/m Cu and 30 % m/m Zn) were obtained from baths that contained allyl alcohol as an additive.

The composition of the electrodeposition bath has been widely studied, in combination with the effects of pH, current density (or applied potential), temperature and stirring speed, as a means of controlling the coating composition and properties. Takata & Sumodjo (2007) have observed the effects of pH and the $[Co^{2+}]/[Pd^{2+}]$ ratio on the composition, morphology and magnetic properties of electrodeposited Co–Pd thin films from a glycine bath. Initially, the pH was kept constant while the $[Co^{2+}]/[Pd^{2+}]$ ratio was varied. As a result, the current efficiency was not affected, whereas the alloy composition changed, which resulted in an increase in the Pd content in the coating. Under these conditions, cracks were always observed, and increasing palladium content in the Co–Pd alloy resulted in deposits with more cracks. The magnetic properties also changed with the increase in Pd content in the coating. Then, the authors maintained the bath composition and varied the pH between 6.5 and 9.4, which did not change the current efficiency or coating composition. All morphological and magnetic changes were related to other parameters (in this case, current density). While investigating the electrodeposition of hybrid coatings, Wu et al. (2004) have demonstrated the influence of the plating parameters on the electrodeposition of a Co–Ni–Al_2O_3 composite from a sulfamate electrolyte. Similar to the methods in previously discussed investigations, certain parameters were kept constant (such as the solution pH, deposition temperature, and current density) while others were varied separately (the cobalt concentration and the alumina concentration in the bath, for example). The authors were able to determine the best conditions to produce composite coatings with the maximum alumina content.

The current characteristics also affect the coating composition and morphology and need to be controlled to produce coatings with the desired properties. Vicenzo et al. (2010) have compared the tin deposition process from a commercial bath using direct and pulsed current. The authors noted several differences in the microstructure and in the properties of the coating after the application of pulsed current. Moreover, changes in the duty cycle and in the frequency of the pulsed current deposition mode also affected the cathodic efficiency and the coating microstructure. Ramanauskas et al. (2008) have studied the deposition of Zn–Ni, Zn–Co, and Zn–Fe alloys using pulsed current mode. The authors performed the electrodeposition experiments by varying one parameter at a time, with the other parameters being fixed at standard levels. The plating parameters studied included the peak cathodic current, i_p, the time of current on, t_{on}, and the time of current off, t_{off}. The coating compositions—mainly the zinc content in the studied alloy coatings—were strongly affected by these parameters. Topographic and phase changes were also noted. The best parameters for each alloy coating deposition were determined, and the coatings were produced on a steel substrate under their respective optimum conditions. The polarization curves of these coating/substrate systems were obtained in a naturally aerated NaCl + $NaHCO_3$ solution at pH = 6.8. The lowest i_{corr} values under all deposition conditions were exhibited by Zn–Ni coatings, whereas Zn–Co and Zn–Fe, except for t_{off} variation, exhibited similar corrosion resistance.

The previously discussed works and several other studies found in the literature produced valid and consistent results that can contribute to a better comprehension of the electrodeposition process and the mechanisms involved in the reduction of one or more ions in an aqueous bath. However, the use of univariate methods cannot allow the evaluation of simultaneous effects of more than one deposition parameter on the same variable. This limitation means that the joint effects of simultaneous variations in bath composition, current density, and temperature, for example, cannot be obtained by these methods. Consequently, the optimum conditions determined by variation of one of these parameters at time may not always represent the real process.

3. Optimization strategies and multivariate approach as a means for enhancing electrodeposition processes

Optimization strategies are procedures followed when, for instance, a product or a process is to be optimized. In an optimization, one tries to determine the optimal settings or conditions for a number of factors. Factors are parameters than can be set and reset at given

Experimental Design and Response Surface Analysis as Available Tools for Statistical Modeling and Optimization
of Electrodeposition Processes

153

levels, e.g., temperature, pH, reagent concentrations, reaction time, etc., and that affect the responses of a product or process. The factors and their level ranges form the experimental domain within which the global optimum is sought. Factors also might "interact." For instance, a two-factor interaction occurs when the influence of one factor on the response is different at different levels of the second factor.

When only one factor needs to be optimized, a simple univariate procedure is performed. Nonetheless, two or more factors are typically studied using either univariate or multivariate optimization strategies (Massart et al., 1997).

An applied univariate procedure, as previously discussed, is the one-variable-at-a-time (OVAT) approach, where only one factor at a time is varied and optimized. The OVAT procedure, however, has some disadvantages: interactions between factors are not taken into consideration; many experiments are needed when the number of factors increases; only a small part of the experimental domain is analyzed; the global optimum might not be found; and the found optimal conditions might depend on the starting conditions (Massart et al., 1997).

However, a multivariate approach varies several factors simultaneously. Multivariate approaches are subdivided into sequential and simultaneous procedures (Massart et al., 1997). Sequential procedures involve a few initial experiments and use their results to define the subsequent experiment (s) (Dejaegher & Vander Heyden, 2009). Sequential approaches can be applied when the experimental domain containing the optimum is *a priori* unknown, but are limited to the optimization of only one response. Simultaneous procedures perform a predefined number of experiments according to a well-defined experimental set-up, e.g., an experimental design (Massart et al., 1997).

An experimental design is an experimental setup to simultaneously evaluate numerous factors at given number of levels in a predefined number of experiments. Rigorously, experimental designs can be divided into screening designs and response surface designs. Screening designs involve screening a relatively large number of factors in a relatively small number of experiments. They are used to identify the factors with the strongest influence. Typically, the factors are evaluated at two levels in these designs. Response surface designs are used to find the optimal levels of the most important factors (which are sometimes selected from a screening design approach). In these designs, factors are examined at a minimum of three levels. The optimal conditions are usually derived from response surfaces build with the design results.

The optimization of a method is often divided into screening and optimization phases. During the screening phase, all factors that potentially influence the responses of interest are tested to indicate those with the largest effects. These most important factors are then further explored in the optimization phase, where their best settings, i.e., the optimum conditions, are determined. Screening and response surface designs, respectively, are applied in these steps (Massart et al., 1997).

After optimization, the method should be validated, i.e., evaluated whether it can be applied for its intended purpose (s). One of the method validation items is robustness testing, which evaluates the effects of small changes in the factors on the considered responses and which applies screening designs for this purpose (Dejaegher & Vander Heyden, 2009; Montgomery, 2005).

The classic screening and response surface designs are now presented. Some applications in electroplating processes will be further discussed in the context of process optimization.

3.1 Screening designs

Screening designs are used to indicate the most important factors from those that potentially influence the considered responses. Screening designs are applied in the context of optimizing separation techniques during screening and in robustness testing, and in the context of optimizing processes. Most often, two-level screening designs, such as fractional factorial design, are used (Montgomery, 2005), which allow the examination of a relatively large number of factors f at L = 2 levels in a relatively small number of experiments ($N \geq f + 1$). When f is small, two-level full factorial designs might also be applied for screening purposes (Montgomery, 2005). These designs allow the simultaneous investigation of qualitative and quantitative factors.

3.2 Two-level full factorial designs

A two-level full factorial design contains all possible combinations between the f factors and their L = 2 levels, leading to $N = L^f = 2^f$ experiments to be performed. Suppose that three factors, A, B, and C, each at two levels, are of interest. The design is called a 2^3 factorial design, and the eight treatment combinations are shown in Table 1. Using the "-1 and 1" notation to represent the low and high levels of the factors, we list the eight runs in the 2^3 design as in Table 1. This is sometimes called the design matrix. These designs allow the estimation of all main (i.e., of the factors) and interaction effects between the considered factors (Montgomery, 2005). The interaction effects are calculated from the columns of contrast coefficients (Table 2).

Run no.	Factors		
	A	B	C
1	-1	-1	-1
2	1	-1	-1
3	-1	1	-1
4	1	1	-1
5	-1	-1	1
6	1	-1	1
7	-1	1	1
8	1	1	1

Table 1. The 2^3 two-level full factorial design for 3 factors

3.3 Two-level fractional factorial designs

A two-level fractional factorial 2^{f-v} design contains a fraction of the full factorial design, and allows examining f factors at two levels in $N = 2^{f-v}$ runs, with $1/2^v$ representing the fraction of the full factorial ($v = 1, 2, 3, \ldots$) (Table 3) (Montgomery, 2005). For fractional factorial designs, N is a power of ($N = 8, 16, 32, \ldots$). Because only a fraction of a full factorial design is carried out, some information is lost.

Run no.	Contrast coefficients			
	AB	AC	BC	ABC
1	1	1	1	-1
2	-1	-1	1	1
3	-1	1	-1	1
4	1	-1	-1	-1
5	1	-1	-1	1
6	-1	1	-1	-1
7	-1	-1	1	-1
8	1	1	1	1

Table 2. The columns of contrast coefficients for the interactions

It is important to note that not all main and interaction effects can be estimated separately anymore. Some effects are confounded, meaning that they are estimated together (Table 3). For instance, in a half-fraction factorial design, each estimated effect is a confounding of two effects.

Run no.	Factors				Contrast coefficients		
	A	B	C	D	I_1	I_2	I_3
1	-1	-1	-1	-1	1	1	1
2	1	-1	-1	1	-1	-1	1
3	-1	1	-1	1	-1	1	-1
4	1	1	-1	-1	1	-1	-1
5	-1	-1	1	1	1	-1	-1
6	1	-1	1	-1	-1	1	-1
7	-1	1	1	-1	-1	-1	1
8	1	1	1	1	1	1	1

Table 3. The 2^{4-1} fractional factorial design for 4 factors, and the columns of contrast coefficients that can still be constructed. A = BCD, B = ACD, C = ABD, D = ABC, I_1 = AB + CD, I_2 = AC + BD, I_3 = AD + BC.

3.4 Supersaturated designs as screening designs

A saturated design is defined as a fractional factorial in which the number of factors or design variables k is given by $k = N - 1$, where N is the number of runs. Recently, considerable interest has been shown in the development and use of a supersaturated design for factor screening experiments. In this design, the number of variables k is $k > N - 1$, and these designs usually contain more variables than runs (Montgomery, 2005).

Supersaturated designs have not found popular use. They are, however, an interesting and potentially useful method for experimentation with systems that involve many variables and where only a very few of these variables are expected to produce large effects (Montgomery, 2005). Their application in the screening design for electroplating processes has not yet been reported.

3.5 Response surface designs

The most important factors—either found from screening or known from experience—are investigated in more detail using response surface designs. These, in fact, are used to determine the optimal conditions for the factors. In these designs, only quantitative and mixture-related factors are examined because the responses considered are modeled as a function of the factors. The response surfaces are then visualized. Most often, only two or three factors are further explored. For more than two factors, only fractions of the entire response surface are visualized (Massart et al., 1997; Montgomery, 2005).

Response surface methodology, by definition, is a collection of mathematical and statistical techniques that are useful for the modeling and analysis of problems in which a response of interest is influenced by many variables; the aim is to optimize this response (Montgomery, 2005). For example, suppose that a plating engineer wishes to find the levels of current density (x_1) and stirring rate (x_2) that maximize the yield (Y) of a process. The process yield is a function of the levels of current density and stirring rate, such as (Equation 1)

$$y = f(x_1, x_2) + \varepsilon \tag{1}$$

where ε represents the error observed in the response y. If we denote the expected response by (Equation 2)

$$E(y) = f(x_1, x_2) = \eta \tag{2}$$

then the surface represented by $\eta = f(x_1, x_2)$ is called a response surface.

The response surface designs can be divided into symmetrical and asymmetrical designs, depending on their appropriateness for use in an asymmetrical domain (Massart et al., 1997; Montgomery, 2005).

3.6 Symmetrical designs

Symmetrical designs cover a symmetrical experimental domain. They contain, for example, three-level full factorial and central composite designs. Often, in these designs, the central point is replicated 3–5 times, usually to estimate the experimental error. A three-level full factorial design contains all possible combinations between the f factors and their $L = 3$ levels, which leads to $N = L^f = 3^f$ runs, including one central point. Thus, for two factors, 9 runs are needed, whereas the inclusion of three factors requires 27 runs (Montgomery, 2005).

A central composite design (CCD) contains a two-level full factorial design (2^f runs), a star design (2^f experiments) and a central point, which requires $N = 2^f + 2^f + 1$ runs to investigate f factors (Massart et al., 1997; Montgomery, 2005). Thus, for two factors, 9 runs are needed, whereas for three factors, 15 are needed. The points of the full factorial design are situated at the factor levels −1 and +1, those of the star design are situated at the factor levels 0, −α and +α, and the central point at the factor levels is situated at 0. Depending on the α value, two CCDs exist, i.e., a face-centered CCD (FCCD) with $|\alpha| = 1$, which examines the factors at three levels, and a circumscribed CCD (CCCD) with $|\alpha| > 1$, which examines the factors at five levels. For a so-called rotatable CCCD, the α level should be $|\alpha| = (2^f)^{1/4}$, i.e., 1.41 and 1.68 for 2 and 3 factors, respectively (Massart et al., 1997; Montgomery, 2005).

A Box–Behnken design contains N = (2f (f − 1)) + 1 runs, of which one is the center point (Montgomery, 2005). For two factors, no design is described. For three factors, 13 runs are described to be performed (Table 4).

Run no.	Factors		
	A	B	C
1	1	1	0
2	1	-1	0
3	-1	1	0
4	-1	-1	0
5	1	0	1
6	1	0	-1
7	-1	0	1
8	-1	0	-1
9	0	1	1
10	0	1	-1
11	0	-1	1
12	0	-1	-1
13	0	0	0

Table 4. Box–Behnken design for 3 factors

A Doehlert (uniform shell) design has equal distances between all neighboring runs (Doehlert, 1970). The Doehlert design for two factors consists of the six vertices of a hexagon with a center point, which requires N = 7 experiments. The design for three factors consists of a centered dodecahedron, which requires N = 13 runs (Table 5). Contrary to the aforementioned response surface designs, the factors are varied at different numbers of levels, e.g., one at three and one at five levels in the two-factor design, and one at three, one at five, and one at seven levels in the three-factor design.

Run no.	Factors		
	A	B	C
1	1	0	0
2	0.5	0.866	0
3	0.5	0.289	0.816
4	-1	0	0
5	-0.5	-0.866	0
6	0.5	0.289	0.816
7	0.5	-0.866	0
8	0.5	-0.289	-0.816
9	0	0.577	-0.816
10	-0.5	0.866	0
11	-0.5	0.289	0.816
12	0	0.577	0.816
13	0	0	0

Table 5. Doehlert designs for three factors

3.7 Asymmetrical designs

When an asymmetrical domain should be investigated, asymmetrical designs, such as designs constructed with the uniform mapping algorithm of Kennard and Stone, can be applied (Massart et al., 1997; Kennard, 1969). These designs are called asymmetrical because when their experiments are plotted, they result in an asymmetrical shape when an asymmetrical domain is studied. These designs can also be used in a symmetrical domain, and then a symmetric shape may be obtained. Asymmetric designs are used because symmetric designs in an asymmetric domain are problematic: either they are too large and require experiments in an impossible area, or they are too small and a considerable part of the domain is not covered (Massart et al., 1997; Montgomery, 2005).

Most often, from a response surface design, the general model build for f factors is (Equation 3)

$$y = \beta_0 + \sum_{i=1}^{f} \beta_i x_i + \sum_{i=1}^{f} \beta_{ij} x_i x_j + \sum_{i=1}^{f} \beta_{ii} x_i^2 \tag{3}$$

where y is the response, β_0 is the intercept, β_i are the main coefficients, β_{ij} are the two-factor interaction coefficients, and β_{ii} are the quadratic coefficients. Occasionally, the interaction terms are restricted to two-factor interactions $(x_i x_j)$ and the higher-order interactions are neglected, as in Equation 3. Occasionally, the non-significant terms of the model are deleted after a statistical analysis.

3.8 Data interpretation

3.8.1 Screening designs

From the results of a full factorial or fractional factorial design, the effect of each factor X on each response Y is estimated as shown in Equation 4:

$$E_X = \frac{\text{contrast}}{n2^{k-1}} \tag{4}$$

where E_x represents the effect or interaction, n is the number of replicates and k is the number of variables or factors.

Generally, a regression model estimates the relation between N x 1 response vector y, and the N x t model matrix X, where N is the number of design runs, and t is the numbers of terms included in the model. The model matrix is obtained by adding a column of ones before the $t - 1$ design matrix columns, which consists of coded factor levels and the column of contrast coefficients, as defined by the experimental design (Equation 5):

$$y = (X\beta) + \varepsilon \tag{5}$$

where β is the t x 1 vector regression coefficients and ε is an error vector. The regression coefficients, b, are calculated using least squares regression (Equation 6):

$$b = (X^T X)^{-1} X^T y \tag{6}$$

where X^T is the transposed matrix of X.

Because effects estimate the change in response when the factor levels are changed from -1 to $+1$ and the coefficients are between levels 0 and $+1$, both are related as follows (Equation 7):

$$E_X = 2b_X \tag{7}$$

Usually, a graphical and/or statistical interpretation of the estimated effects is performed to determine their significance. Graphically, normal probability or half-normal probability plots can be drawn (Montgomery, 2005). In these plots, the unimportant effects are found on a straight line through zero, whereas the important effects deviate from this line.

The statistical interpretations usually involve a calculation of a *t-test* statistic for the factors (Equation 8) and a comparison between this *t-value* and a limit value, $t_{critical}$, or between the effect E_X and a critical effect, $E_{critical}$, respectively. All effects that are greater than or equal to this $E_{critical}$ in absolute value are then considered significant (Montgomery, 2005):

$$t = \frac{|E_X|}{(SE)_e} \Leftrightarrow t_{critical} \tag{8}$$

with $(SE)_e$ being the standard error of an effect. The critical t-value, $t_{critical}$, depends on the number of degrees of freedom associated with $(SE)_e$ and on the significance level, usually $p = 0.05$. The critical effect, $E_{critical}$, is then obtained as follows (Equation 9):

$$E_{critical} = t_{critical}(SE)_e \leftrightarrow |E_X| \tag{9}$$

The standard error of an effect, $(SE)_e$, can be estimated from different data: from the variance of replicated runs, from *a priori* declared negligible effects, or from *a posteriori* defined negligible effects. We consider the last two approaches the most appropriate for the proper estimation of $(SE)_e$.

3.8.2 Optimization using the desirability function

Many response surfaces imply the analysis of several responses. The simultaneous consideration of multiple responses involves first building an appropriate response surface model for each response and then attempting to find a set of operating conditions that, in some sense, optimizes all responses or at least keeps them within desired ranges. A useful approach to the optimization of multiple responses is the use of the simultaneous optimization technique popularized by Derringer and Suich (Derringer & Suich, 1980). Their procedure makes use of desirability functions. The general approach is to first convert each response Y_i, into an individual desirability function d_i that varies over the range $0 \le d_i \le 1$, where if response Y_i is at its goal or target, then $d_i = 1$, and if the response is outside an acceptable region, $d_i = 0$. Next, the design variables are chosen to maximize the overall desirability, $D = (d_1 \times d_2 \times ... \times d_m)^{1/m}$, where there are m responses (Ferreira et al., 2007).

4. The main results presented in the literature concerning the use of statistic methods in electrodeposition

In the 1990s, optimization started to be essential for development of cheap and rapid process technologies. The need for more refined and functional materials, led by the electronics industry, increased the requirements for process control in electrodeposition and stimulated the use of statistical approaches for the optimization of deposition parameters. Furthermore, experimental designs had already found widespread applications in the physical sciences, mainly because of the development of sophisticated and dedicated software for statistical analyses and numerical simulations. Another factor that contributed to the increase in research using experimental design was the extensive and world-wide use of personal computers with Internet access, mainly at the end of the 1990s. Since then, the number of works found in the literature concerning experimental design, response surface methodology, and other statistical evaluations in electrodeposition processes increased considerably. These works include the evaluation of several deposition parameters in the coatings' properties and/or in their formation mechanisms. The statistical packages of well-known software and dedicated statistics programs are now used, which also enhance the presentation of the results and the response surface designs diagrams.

In the context of optimizing electrodeposition processes, the application of screening and response surface designs has already been discussed and reviewed frequently (Bezerra et al., 2008; Ferreira et al., 2004). For screening, primarily two-level full factorial (Santana et al., 2006; 2007a; 2007b; Souza e Silva et al., 2006) and fractional factorial designs (Wery et al., 1999; Hu & Bay, 2001a; 2001b; Tsay & Hu, 2002; Hu et al., 2003; Bai & Hu, 2005; Hu et al., 2006; Dubent et al., 2007) have been applied. For the actual optimization of important factors, response surface designs, such as central composite (San Martín et al., 1998; Hu & Bay, 2001a; 2001b; Tsay & Hu, 2002; Hu et al., 2003; Bai & Hu, 2005; Hu et al., 2006; Silva et al., 2008; Musa et al., 2008; Poroch-Seritana et al., 2011), Box–Behnken (Morawej et al., 2006), and Doehlert (Chalumeau et al., 2004; 2006], have been used.

4.1 Classic applications concerning the use of screening and optimization in electrodeposition

The group of Chi-Chang Hu has presented some electrodeposition experiments. To optimize the hydrogen evolution activity of Zn–Ni and Ni–P deposits, respectively, different experimental strategies, including the fractional factorial design (FFD), the path of the steepest ascent study, and the central composite design (CCD) coupled with the response surface methodology (RSM), were adopted (Hu & Bay, 2001a; 2001b; Hu et al., 2003). The same experimental strategies were used to find the optimal plating conditions in the pulse-reverse electroplating mode for non-anomalous plating of Co–Ni and Fe–Ni deposits, from chloride solutions (Bai & Hu, 2005; Tsay & Hu, 2002). The same group again reported the aforementioned strategies to study the electroplating conditions of a direct-current (DC) plating mode for the co-deposition of Sn–Zn deposits, and their composition close to the eutectic point were achieved from chloride solutions. The temperature of the plating bath, pH, and the metallic ion ratio (i.e., Sn^{4+}/Zn^{2+} ratio) were found to be the key factors that affect the composition of Sn–Zn deposits in the fractional factorial design study. The effects of pH and the temperature of the plating solution on the composition of Sn–Zn deposits were examined using a regression model in the central composite design study (Hu et al., 2006).

Concerning the electrodeposition of ternary alloys, the Prasad group has reported some studies. Ternary alloys Ni–W–B, Ni–Fe–Mo, and Ni–W–Co were electrodeposited, and operational parameters related to their corrosion resistance and deposition efficiency were optimized (Santana et al., 2006; Santana et al., 2007a; Santana et al., 2007b). The Doehlert experimental design was applied to study Au–Co electroplating optimization, and validation was performed by means of statistical analysis (Chalumeau et al., 2004; 2006). Fractional factorial design was used to analyze the influence of plating conditions on the cathode efficiency of zinc barrel electroplating and on the quality of deposited layers for low-cyanide electrolytes (Wery et al., 1999), as well as to study the determination of the suitable electroplating conditions (i.e., electrolyte composition and cathode current density) to produce 70Sn–30Zn electrodeposits (Dubent et al., 2007).

Experimental design has been applied to study the influence of several parameters on the electroplating of Cu–Zn alloy in cyanide medium (Musa et al., 2008), and response surface modeling and optimization has been used by the Senna group to study the influence of deposition parameters on the electrodeposition of both Cu–Zn and Cu–Co alloys from in citrate medium (Ferreira et al., 2007; Silva et al., 2008). Cembero & Busquets-Mataix (2009) have also used experimental design to study the most important parameters in the production of ZnO crystals by electrodeposition.

The central composite experimental design and response surface methodology have been employed for statistical modeling and analysis of results dealing with nickel electroplating process. The empirical models developed in terms of design variables (current density J (A/dm^2), temperature T (°C) and pH) have been found to be statistically adequate to describe the process responses, i.e., cathode efficiency Y (%), coating thickness U (µm), brightness V (%) and hardness W (HV). The multi-response optimization of the nickel electroplating process has been performed using the desirability function approach (Poroch-Seritana et al., 2011). Ferreira et al. (2007), studying the electrodeposition process of Cu–Zn alloys from citrate bath, also used multi-response optimization and the desirability function approach to determine the best conditions (stirring speed and current density values) to deposit a Cu–Zn alloy coating with the best anticorrosive performance. Silva et al. (2007) were also able to obtain the best conditions to produce Cu–Co alloys from citrate baths, using the desirability function approach. Unfortunately, there are still few works concerning this topic in the literature.

5. Conclusions

This chapter has provided an overview of both classic and advanced experimental design setups and their data interpretation. Statistical methods, such as experimental design and response surface methodology, could be used to help the plating engineer find a better relationship among all the parameters that might simultaneously influence the properties and performance of metallic/alloy or hybrid coatings. Rather uncommon experimental setups, such as the Doehlert design as screening design, were also considered. This was followed by a discussion of the applications using fractional factorial designs as screening designs and also of the application of desirability functions to solve problems of multi-responses. This last topic may be very useful for plating engineers to produce more adequate coatings for desired applications.

6. Acknowledgements

The authors would like to thank the Brazilian Council of Research and Development (CNPq), the Rio de Janeiro State Foundation for Research (FAPERJ), and the Rio de Janeiro State University – Prociência Program for research grants.

7. References

Bai, A. & Hu, C. C. (2005) Composition controlling of Co–Ni and Fe–Co alloys using pulse-
    Bai, A. & Hu, C. C. (2005) Composition controlling of Co–Ni and Fe–Co alloys using pulse-reverse electroplating through means of experimental strategies. *Electrochimica Acta*, Vol. 50, No. 6, (January 2005), pp. 1335–1345, ISSN: 0013-4686.
Bezerra, M. A.; Santelli, R. E.; Oliviera, E. P.; Villar, L. S. & Escaleira, L. A. (2008) Response surface methodology (RSM) as a tool for optimization in analytical chemistry. *Talanta*, Vol. 76, No. 5 (September 2008), pp. 965–977. ISSN: 0039-9140
Bozzini, B.; Mele, C.; D'Urzo, L.; Giovannelli, G. & Natali, S. Electrodeposition of Cu from acidic sulphate solutions in the presence of PEG: an electrochemical and spectroelectrochemical investigation – Part I. *Journal of Applied Electrochemistry*, Vol. 36, No. 1 (January 2006), pp. 87-97. ISSN: 0021-891X.
Brankovic, S. R.; George, J.; Bae, S. -E. & Litvinova, D. (2009) Critical Parameters of Solution Design for Electrodeposition of 2.4 T CoFe Alloys. *The Electrochemical Society Transactions*, Vol. 16, No. 45, (October 2009), pp. 75-87. ISSN: 1938-5862.
Cembrero, J.& Busquets-Mataix, D. (2009) ZnO crystals obtained by electrodeposition: Statistical analysis of most important process variables. *Thin Solid Films*, Vol. 517, No. 9, (March 2009), pp. 2859-2864. ISSN: 0040-6090.
Chalumeau, L.; Wery, M.; Ayedi, H. F.; Chabouni, M. M. & Leclere, C. (2004) Application of a Doehlert experimental design to the optimization of an Au–Co plating electrolyte. *Journal of Applied Electrochemistry* Vol. 34, No. 11 (December 2004), pp. 1177–1184. ISSN: 0021-891X.
Chalumeau, L.; Wery, M.; Ayedi, H. F.; Chabouni, M. M. & Leclere, C. (2006) Development of a new electroplating solution for electrodeposition of Au–Co alloys. *Surface and Coatings Technology*, Vol. 201, No. 3/4 (October 2006), pp. 1363 – 1372. ISSN: 0257-8972.
Derringer, G. & Suich, R. (1980) Simultaneous optimization of several response variables. *Journal of Quality Technology*, Vol. 12, No. 4 (October 1980), pp. 214-219. ISSN: 0022-4065
Dejaegher, B. & Heyden, Y. V. (2009) Sequential optimization methods. In: *Comprehensive Chemometrics, vol. 1*, S. Brown; R. Tauler & B. Walczak, (Eds.), 547–575, Elsevier, ISBN: 978-0444527035, Oxford, Great Britain.
Drissi-Daoudi, R.; Irhzo, A. & Darchen, A. (2003) Electrochemical investigations of copper behaviour in different cupric complex solutions: Voltammetric study. *Journal of Applied Electrochemistry*, Vol. 33, No. 4 (April 2003), pp. 339–343. ISSN: 0021-891X
Doehlert, D.H. (1970) Uniform shell designs. *The Royal Statistical Society Series C-Applied Statistics*, Vol. 19, No. 3 (March 1970), pp. 231–239. ISSN: 0035-9254.

Experimental Design and Response Surface Analysis as Available Tools for Statistical Modeling and Optimization
of Electrodeposition Processes

163

Düzgün, E.; Ergun, Ü.; Ercan, F.; Sert, M.; Aksu, M. L. & Atakol, O. (2010) Preparation of thin film alloys with electrodeposition from heterotrinuclear M1–M2–M1 complexes (M1: Ni^{II}, Cu^{II}; M2: Cu^{II}, Co^{II}, Zn^{II}, Cd^{II}). *Journal of Coating Technology Research*, Vol. 7, No. 3, (May 2010), pp. 365–371. ISSN: 1547-0091.

Dubent, S.; de Petris-Wery, M.; Saurat, M. & . Ayedi, H. F. (2007) Composition control of tin–zinc electrodeposits through means of experimental strategies. *Materials Chemistry and Physics*, Vol. 104, No. 1 (July 2007), pp. 146–152. ISSN: 0254-0584.

Ferreira, S. L. C.; dos Santos, W. N. L.; Quintella, C. M.; Neto, B. B. & Bosque-Sendra, J. M. (2004) Doehlert matrix: a chemometric tool for analytical chemistry — review. *Talanta*, Vol. 63, No. 4 (July 2004), pp. 1061–1067. ISSN: 0039-9140

Ferreira, F. B. A.; Silva, F. L. G.; Luna, A. S.; Lago, D. C. B. & Senna, L. F. (2007) Response surface modeling and optimization to study the influence of deposition parameters on the electrodeposition of Cu–Zn alloys in citrate medium. *Journal of Applied Electrochemistry*, Vol. 37, No. 4, (April, 2007), pp. 473–481. ISSN: 0021-891X

Finot, M. O.; Braybrook, G. D.; McDermott, M. T. (1999) Characterization of electrochemically deposited gold nanocrystals on glassy carbon electrodes. *Journal of Electroanalytical Chemistry*, Vol. 466, No. 2 (May 1999), pp. 234–241. ISSN: 1572-6657.

Gomez, E. & Valles, E. (1999) Electrodeposition of Co plus Ni alloys on modified silicon substrates. *Journal of Applied Electrochemistry*, Vol. 29, No. 7 (July 1999), pp. 805-812. ISSN: 0021-891X

Gupta, M. & Podlaha, E. J. (2010) Electrodeposition of CuNiW alloys: thin films, nanostructured multilayers and nanowires. *Journal of Applied Electrochemistry*, Vol. 40, No. 7 (July 2010), pp. 1429–1439. ISSN: 0021-891X

Gurrappa, I. & Binder, L. (2008) Electrodeposition of nanostructured coatings and their characterization — a review. *Science and Technology of Advanced Materials*, Vol. 9, No. 4 (December 2008), pp. 1-11. ISSN: 1468-6996.

Healy, J. P.; Pletcher, D. & Goodenough, M. (1992) The chemistry of the additives in an acid copper electroplating bath Part I. Polyethylene glycol and chloride ion. *Journal of Electroanalitical Chemistry*, V. 338, No. 1/2 (October 1992), pp. 155-165. ISSN: 1572-6657.

Healy, J. P.; Pletcher, D. & Goodenough, M. (1992) The chemistry of the additives in an acid copper electroplating bath Part II. The instability of 4,5-dithiaoctane-l, & disulphonic acid in the bath on open circuit. *Journal of Electroanalitical Chemistry*, Vol. 338, No. 1/2 (October 1992), pp. 167-177. ISSN: 1572-6657.

Hu, C. C. & Weng, C. Y. (2000) Hydrogen evolving activity on nickel-molybdenum deposits using experimental strategies. *Journal of Applied Electrochemistry*, Vol. 30, No. 4 (April 2000), pp. 499-506. ISSN: 0021-891X.

Hu, C. C. & Bai, A. (2001) Optimization of hydrogen evolving activity on nickel-phosphorous deposits using experimental strategies. *Journal of Applied Electrochemistry*, Vol. 31, No. 5 (May 2001), pp. 565-572. ISSN: 0021-891X

Hu, C. C. & Bai, A. (2001) Composition control of electroplated nickel-phosphorus deposits. *Surface and Coatings Technology*, Vol. 137, No. 2-3 (March 2001), pp. 181-187. ISSN: 0257-8972.

Hu, C. C.; Tsay, C. H. & Bai, A. (2003) Optimization of the hydrogen evolution activity on zinc/nickel deposits using experimental strategies. *Electrochimica Acta*, Vol. 48, No. 7 (Feruary 2003), pp. 907_918. ISSN: 0013-4686.

Hu, C. C.; Wang, C. K. & Lee, G. L. (2006) Composition control of tin–zinc deposits using experimental strategies. . *Electrochimica Acta*, Vol. 51, No. 18 (May 2006), pp. 3692–3698. ISSN: 0013-4686.

Huang, H. & Yanga, X. (2005) One-step, shape control synthesis of gold nanoparticles stabilized by 3-thiopheneacetic acid. *Colloids and Surfaces A: Physicochemistry and Engineering Aspects*, Vol. 255, No. 1/3 (March 2005), pp. 11–17. ISSN: 0927-7757.

Illy, B.; Shollock, B. A.; MacManus-Driscoll, J. L. & Ryan, M. P. (2005) Electrochemical growth of ZnO Nanoplates. *Nanotechnology*, Vol. 16, No. 2 (January 2005), pp. 320–324. ISSN: 0957-4484.

Jeerage, K. M. & Schwartz, D. T. (2004) Alloy Electrodeposition. *The Electrochemical Society Interface*. Vol. 16, No. 1 (January 2004), pp. 23 - 24. ISSN: 1064-8208.

Kennard, R. W. & Stone, L. A. (1969) Computer aided design of experiments.*Technometrics*, Vol. 11, No. 1 (January 1969), pp. 137–148. ISSN: 0040-1706.

Kharlamov, V. I.; Kruglikov, S. S.; Grigoryan, N. S. & Vagramyan, T. A. (2001) Microdistributions of electrolytic alloys and their components. *Russian Journal of Electrochemistry*, Vol. 37, No. 7 (July 2001), pp. 661–669. ISSN: 1023-1935.

Kim, H.; Moon, J. Y. & Lee, H. S. (2009) Growth of ZnO nanorods on various substrates by electrodeposition. *Electronic Materials Letters*, Vol. 5, No. 3, (September 2009), pp. 135-138. ISSN: 1738-8090.

Landolt, D. (1994) Electrochemical and materials science: aspects of alloy deposition. *Electrochimica Acta*, Vol. 39, No. 8/9, (June 1994), pp. 1075- 1090. ISSN: 0013-4686.

Lu, D. L. & Tanaka, K.I. (1996) Gold Particles Deposited on Electrodes in Salt Solutions under Different Potentials. *Journal of Physical Chemistry*, Vol. 100, No. 5 (February 1996), pp. 1833-1837. ISSN: 1089-5639.

Massart, D. L.; Vandeginste, B. G. M.; Buydens, L. M. C.; de Jong, S. P.; Lewi, J.& Smeyers-Verbeke, J. (1997) *Handbook of chemometrics and qualimetrics: Part A*, Elsevier, ISBN: 0-444-89724-0,Amsterdam, Netherlands.

Mohanty, U. S. (2011) Electrodeposition: a versatile and inexpensive tool for the synthesis of nanoparticles, nanorods, nanowires, and nanoclusters of metals. *Journal of Applied Electrochemistry*, Vol. 41, No. 3 (March 2011), pp. 257–270. ISSN: 0021-891X

Montgomery, D.C. (2005) *Design and analysis of experiments, 6th edition*, John Wiley, ISBN: 0-471-48735X, New York, USA.

Morawej, N.; Ivey, D. G. & Akhlaghi, S. (2006) Improvements in the process for electrodeposition of Au-Sn alloys, *Proceedings of The International Conference on Compound Semiconductor Manufacturing Technology - CS MANTECH Conference*, pp. 201-204, Vancouver, British Columbia, Canada, April 24-27, 2006. Available from http://www.csmantech.org/Digests/2006/index2006.html#.

Mosavat, S. H.; Bahrololoom, M. E. & Shariat, M.H. (2011) Electrodeposition of nanocrystalline Zn–Ni alloy from alkaline glycinate bath containing saccharin as additive. *Applied Surface Science*, vol. 257, No. 20 (August 2011), pp. 8311– 8316. ISSN: 0169-4332

Musa, A. Y. Q.; Slaiman, J. M. A.; Kadhum, A. H. & Takriff, M. S . (2008) Effects of agitation, current density and cyanide concentration on Cu-Zn alloy electroplating. *European Journal of Scientific Research*, Vol. 22, No. 4 (October 2008), pp. 517–524. ISSN: 1450-216X.450-216X

Nayana, K. O.; Venkatesha, T. V.; Praveen, B. M. & Bástala, K. (2011) Synergistic effect of additives on bright nanocrystalline zinc electrodeposition. *Journal of Applied Electrochemistry*, Vol. 41, No. 1 (January, 2011), pp. 39–49. ISSN: 0021-891X

Experimental Design and Response Surface Analysis as Available Tools for Statistical Modeling and Optimization of Electrodeposition Processes

165

Oriňáková, R.; Turoňová, A.; Kladeková, D.; Gálová, M. & Smith, R. (2006) Recent developments in the electrodeposition of nickel and some nickel-based alloys. *Journal of Applied Electrochemistry*, Vol. 36, No. 9 (September 2006), pp. 957–972. ISSN: 0021-891X

Poroch-Seritan, M.; Gutt, S.; Gutt, G.; Cretescu, I.; Cojocaru, C. & Severin, T. (2011) Design of experiments for statistical modeling and multi-response optimization of nickel electroplating process. *Chemical Engineering Research and Design*, Vol. 89, No. 2 (February 2011), pp. 136–147. ISSN: 0263-8762.

Rabiot, D.; Caire, J.-P.; Nguyen, B.; Chainet. E. & Gerard, F. (1998) Optimizing an electrochemical deposition process by use of design of computer experiments. *Analusis*, Vol. 26, No. 8 (October 1998), pp. 281-284. ISSN: 0365-4877.

Ramanauskas, R.; Gudavičiūtė, L.; Ščit, O.; Bučinskienė, D. & Juškėnas, R. (2008) Pulse plating effect on composition and corrosion properties of zinc alloy coatings. *Transactions of the Institute of Metal Finishing*, Vol. 86, No. 2, (March 2008), pp. 103-108. ISSN: 0020-2967.

Roventi, G.; Bellezze, T. & Fratesi, R. (2006) Electrochemical study on the inhibitory effect of the underpotential deposition of zinc on Zn–Co alloy electrodeposition. *Electrochimica Acta*, Vol. 51, No. 13 (March 2006), pp. 2691–2697. ISSN: 0013-4686.

San Martín, V. Sanllorente, S. & Palmero, S. (1998) Optimization of influent factors on nucleation process of copper in solutions containing thiourea using an experimental design. *Electrochimica Acta*, Vol. 44, No. 4 (September 1998), pp. 579-585. ISSN: 0013-4686.

Santana, R. A. C.; Prasad, S.; Campos, A. R. N.; Araújo, F. O.; da Silva, G. P. & de Lima-Neto, P. (2006) Electrodeposition and corrosion behaviour of a Ni–W–B amorphous alloy. *Journal of Applied Electrochemistry*, Vol. 36, No. 1 (January 2006), pp. 105–113. ISSN: 0021-891X

Santana, R. A. C.; Prasad, S.; Moura, E. S.; Campos, A. R. N.; Silva, G. P. & Lima-Neto, P. (2007) Studies on electrodeposition of corrosion resistant Ni–Fe–Mo Alloy. *Journal of Material Science*, Vol 42, No. 7 (February 2007), pp. 2290–2296. ISSN: 0022-2461.

Santana, R. A. C.; Campos, A. R. N.; Medeiros, E. A.; Oliveira, A. L.; Silva, L. M. F. & Prasad, S. (2007) Studies on electrodeposition and corrosion behavior of a Ni–W–Co amorphous alloy. *Journal of Material Science*, Vol 42, No. 22 (July 2007), pp. 9137–9144. ISSN: 0022-2461.

Schwarzacher, W. (2004) Kinetic roughening of electrodeposited films. *Journal of Physics: Condensed Matter*, Vol. 16, No. 26 (July 2004), pp. R859–R880. ISSN: 0953-8984.

Schwartz, M. (1994) Deposition from aqueous solutions: an overview. In: *Handbook of Deposition Technologies for Films and Coatings - Science, Technology and Applications (2nd Edition)*, R.F. Bunshah, (Ed.), 506-616, William Andrew Publishing, London, Great Britain. ISBN: 978-0-8155-1337-7

Senna, L. F.; Díaz, S. L. & Sathler, L. (2003) Electrodeposition of copper–zinc alloys in pyrophosphate-based electrolytes. *Journal of Applied Electrochemistry*, Vol. 33, No. 12 (December 2003), pp. 1155–1161. ISSN: 0021-891X

Senna, L. F.; Díaz, S. L. & Sathler, L. (2005) Hardness Analysis and Morphological Characterization of Copper-Zinc Alloys Produced in Pyrophosphate-Based Electrolytes. *Materials Research*, Vol. 8, No. 3, (July 2005), pp. 275-279. ISSN: 1516-1439.

Silva, F. L. G.; Cruz , V. G. M.; Garcia, J. R.; Luna, A. S.; Lago, D. C. B. & Senna, L. F. (2007) Estudo dos parâmetros para a produção de revestimentos de liga Cu-Co sobre substratos de aço ao carbono com propriedades anticorrosivas. *Proceedings of The 9th International Conference on Equipments Technology – 9th COTEQ*, pp. 1-8, Salvador, Bahia, Brazil, June 12-15, 2007. Available from CD-ROM - http://www.abende.org.br/down2/anais_2011.pdf.

Silva, F. L. G.; Garcia, J. R.; Cruz, V. G. M.; Luna, A. S.; Lago, D. C. B. & Senna, L. F. (2008) Response surface analysis to evaluate the influence of deposition parameters on the electrodeposition of Cu–Co alloys in citrate medium. *Journal of Applied Electrochemistry*, Vol. 38, No. 12 (December 2008), pp. 1763-1769. ISSN: 0021-891X

Skirstymonskaya, B. I. (1964) Electrodeposition of alloys. *Russian Chemical Reviews*, Vol.33 No.4, (April 1964), pp. 221-233. ISSN: 0036-021X

Souza e Silva, P. T.; de Mello, N. T.; Duarte, M. M.; Conceição, M. M.; Montenegro, B. S. M.; Araújo, A. N.; Neto, B. B. & da Silva, V. L. (2006) Nome do artigo. *Journal of Hazard Materials B*, Vol. 128, No. 1 (January 2006), pp. 39–43. ISSN: 0304-3894.

Takata, F. M & Sumodjo, P. T. A. (2007) Electrodeposition of magnetic CoPd thin films: Influence of plating condition. *Electrochimica Acta*, Vol. 52, No. 20 (June 2007), pp. 6089–6096. ISSN: 0013-4686.

Tantavichet, N. & Pritzker, M. (2006) Copper electrodeposition in sulphate solutions in the presence of benzotriazole. *Journal of Applied Electrochemistry*, Vol. 36, No. 1 (January 2006), pp. 49–61. ISSN: 0021-891X

Tsai, T. H.; Thiagarajan, S. & Chen, S. M. (2010) Green synthesized Au–Ag bimetallic nanoparticles modified electrodes for the amperometric detection of hydrogen peroxide. *Journal of Applied Electrochemistry*, Vol. 40, No. 12 (December 2010), pp. 2071–2076. ISSN: 0021-891X

Tsay, P. & Hu, C. C. (2002) Non-anomalous codeposition of iron-nickel alloys using pulse-reverse electroplating through means of experimental strategies. *Journal of The Electrochemical Society*, Vol. 149, No. 10, (August 2002), pp. C492-C497. ISSN: 0013-4651

Vicenzo, A.; Bonelli, S. & Cavallotti, P. L. (2010) Pulse plating of matt tin: effect on properties. *Transactions of the Institute of Metal Finishing*, Vol 88, No 5, (September 2010), pp. 248-255. ISSN: 0020-2967.

Wery, M.; Catonné, J. C.; Ligier, V. & Pagetti, J. (1999) Zinc barrel electroplating using low cyanide electrolytes. *Journal of Applied Electrochemistry*, Vol. 29, No. 6 (June 1999), pp. 733-743. ISSN: 0021-891X

Wua, G.; Lia, N.; Zhoua, D. & Mitsuo, K. (2004) Electrodeposited Co–Ni–Al$_2$O$_3$ composite coatings. *Surface and Coatings Technology*, Vol. 176, No. 1 (January 2004), pp. 157–164. ISSN: 0257-8972.

Yoshida,T.; Zhang, J.; Komatsu, D.; Sawatani, S.; Minoura, H.; Pauporté, T.; Lincot, D.; Oekermann, T.; Schlettwein, D.; Tada, H.; Wöhrle, D.; Funabiki, K.; Matsui, M.; Miura, H. & Yanagi, H. (2009) Electrodeposition of Inorganic/Organic Hybrid Thin Films. *Advanced Functional Materials*, Vol. 19, No. 1 (January 2009), pp. 17–43. ISSN: 1616-3028.

Permissions

The contributors of this book come from diverse backgrounds, making this book a truly international effort. This book will bring forth new frontiers with its revolutionizing research information and detailed analysis of the nascent developments around the world.

We would like to thank Dr. Darwin Sebayang and Prof. Dr. Sulaiman Bin Haji Hasan, for lending their expertise to make the book truly unique. They have played a crucial role in the development of this book. Without their invaluable contribution this book wouldn't have been possible. They have made vital efforts to compile up to date information on the varied aspects of this subject to make this book a valuable addition to the collection of many professionals and students.

This book was conceptualized with the vision of imparting up-to-date information and advanced data in this field. To ensure the same, a matchless editorial board was set up. Every individual on the board went through rigorous rounds of assessment to prove their worth. After which they invested a large part of their time researching and compiling the most relevant data for our readers. Conferences and sessions were held from time to time between the editorial board and the contributing authors to present the data in the most comprehensible form. The editorial team has worked tirelessly to provide valuable and valid information to help people across the globe.

Every chapter published in this book has been scrutinized by our experts. Their significance has been extensively debated. The topics covered herein carry significant findings which will fuel the growth of the discipline. They may even be implemented as practical applications or may be referred to as a beginning point for another development. Chapters in this book were first published by InTech; hereby published with permission under the Creative Commons Attribution License or equivalent.

The editorial board has been involved in producing this book since its inception. They have spent rigorous hours researching and exploring the diverse topics which have resulted in the successful publishing of this book. They have passed on their knowledge of decades through this book. To expedite this challenging task, the publisher supported the team at every step. A small team of assistant editors was also appointed to further simplify the editing procedure and attain best results for the readers.

Our editorial team has been hand-picked from every corner of the world. Their multi-ethnicity adds dynamic inputs to the discussions which result in innovative outcomes. These outcomes are then further discussed with the researchers and contributors who give their valuable feedback and opinion regarding the same. The feedback is then collaborated with the researches and they are edited in a comprehensive manner to aid the understanding of the subject.

Apart from the editorial board, the designing team has also invested a significant amount of their time in understanding the subject and creating the most relevant covers. They scrutinized every image to scout for the most suitable representation of the subject and create an appropriate cover for the book.

The publishing team has been involved in this book since its early stages. They were actively engaged in every process, be it collecting the data, connecting with the contributors or procuring relevant information. The team has been an ardent support to the editorial, designing and production team. Their endless efforts to recruit the best for this project, has resulted in the accomplishment of this book. They are a veteran in the field of academics and their pool of knowledge is as vast as their experience in printing. Their expertise and guidance has proved useful at every step. Their uncompromising quality standards have made this book an exceptional effort. Their encouragement from time to time has been an inspiration for everyone.

The publisher and the editorial board hope that this book will prove to be a valuable piece of knowledge for researchers, students, practitioners and scholars across the globe.

List of Contributors

Misoon Oh and Seok Kim
Pusan National University, South Korea

Darwin Sebayang, Yanuandri Putrasari, Sulaiman Hasan and Mohd Ashraf Othman
Universiti Tun Hussein Onn Malaysia, Malaysia

Pudji Untoro
Badan Tenaga Nuklir Nasional, Indonesia

Frederic Raynal
Alchimer S.A., France

Jae-Hoon Lee
GaN Power Research Group, R&D Institute, Samsung LED Company Ltd., Suwon, Korea

Jung-Hee Lee
Electronic Engineering & Computer Science, Kyungpook National University, Daegu, Korea

Arifa Tahir
Environmental Science Department, LCWU Lahore, Pakistan

Ronaldo Ferreira do. Nascimento, Francisco Wagner de Sousa, Pierre Basílio Almeida Fechine, Paulo de Tarso C. Freire and Marcos Antônio Araujo-Silva
Universidade Federal do Ceara (UFC), Brazil

Vicente Oliveira Sousa Neto
Universidade Estadual do Ceara (UECE-CECITEC), Brazil

Raimundo Nonato Pereira Teixeira
Universidade Regional do Cariri (URCA), Brazil

Lilian F. Senna and Aderval S. Luna
State University of Rio de Janeiro, Chemistry Institute, Pavilhão Haroldo Lisboa da Cunha, Rio de Janeiro, Brazil